Instrumentation Training Course

Volume 2

Electronic Instruments

Second Edition

revised by

Dale R. Patrick, Professor

Department of Industrial Education and Technology
College of Applied Arts and Technology
Eastern Kentucky University

Howard W. Sams & Co., Inc.
4300 WEST 62ND ST. INDIANAPOLIS, INDIANA 46268 USA

Preface

Many instrument people who are competent with mechanical/pneumatic instrumentation are lost on electronic instrumentation. There are, perhaps, two good reasons for this. First, we shy away from electronics because we fear it is "just too complicated." Second, unlike the working of mechanical equipment, which can be seen, electricity cannot be seen, and an elaborate symbolism has resulted which discourages the novice. Much of the symbolism is highly mathematical, which is a further deterrent.

Because of the elaborate symbolism, many have felt that electronics cannot be understood unless the mathematical symbolism is mastered first. Electronics can be mastered in a qualitative way by a direct study of actual industrial instruments without prior study of mathematics or theory. This is possible, provided the course of study is based on functional training methods and an extensive use of the oscilloscope. The oscilloscope will do much toward making it possible to "see" electricity. The only symbolism that must be mastered is the notation of electronic circuitry. The mathematical symbolism for the purposes of this course is covered where it is needed to understand specific circuits.

The method of presenting the material is to plunge directly into real instruments and circuits and to develop the principles and theory when these are required for an understanding of the piece of equipment and circuits under study. This approach has been described as "functional training."

The text is arranged to cover four major aspects of electronic control: measurement circuits and primary devices, transistor electronics, vacuum-tube electronics, and magnetics.

Contents

CHAPTER 1

Measurement Circuits and Primary Devices

INTRODUCTION

The development of electronic instruments that test, measure, and control industrial processes has gone through a rather impressive growth pattern in recent years. It is now common practice for industrial technicians to perform routine measurements using electronic instruments that were considered precision pieces of expensive laboratory equipment only a few years ago. As a result of this, industry in general is very concerned about training its personnel to become competent instrument users in order to take full advantage of the expanding field of electronic instrument technology.

INDUSTRIAL ELECTRONIC SYMBOLS

In the electronics field, a number of symbols are commonly used as a means of representing electrical components in diagrammed form. These symbols serve as the basis for schematic diagrams, which range from simple circuits to very complex circuits. As a rule, only a few basic symbols are in common usage. These symbols, however, are extremely important because they are used many times. It is essential that these symbols, and the components that they represent, be understood in order to effectively interpret the circuitry of a schematic diagram.

We will employ electronic symbols in schematic diagrams throughout this manual. A diagram of an amplifier of a controller is shown in Fig. 1-1 as an example. As you will note in this diagram, the symbols are numbered such as R_{10}, C_2, Q_1, etc. This numbering generally serves as a reference key for replacements.

A number of important component symbols are shown in Fig. 1-2 for reference. These represent some of the more common symbols used in this manual. A more extensive listing of industrial electronic symbols will be found in the Appendix at the back of the manual. Refer to this list when you encounter an unfamiliar symbol.

ELECTRON FLOW

In the study of electronics the generation of electricity is an extremely important concept. In this presentation we will consider only those methods of generation that are particularly applicable to the industrial field. This includes thermocouples, electrochemical cells, and mechanical generators.

It is a well known fact that when two wires of different metals are joined together and placed in a closed loop, as in Fig. 1-3, a current will result. Current simply refers to the flow or movement of electrons.

An electron is one of the basic particles of an atom of matter. An atom, of any one of the 92 natural elements, when divided into its basic particles contains a number of electrons, protons, and neutrons. Fig. 1-4 shows a representative drawing of an atom of carbon. Note that it contains six protons represented by plus signs, six electrons represented by negative signs, and six neutrons represented by the letter N. The center of the atom is called the nucleus. It houses the protons and neutrons. Electrons revolve around the nucleus which tends to attract the electrons by the positive charge of the protons. Atoms are electrically neutral when the number of electrons equals the number of protons in the nucleus. Atoms in general are electrically neutral unless they have been acted upon by some outside source of energy.

The physical makeup of all atoms, as far as we know today, is basically the same as the carbon atom with the only difference being the number of particles. Hydrogen, for example, has 1 electron, 1 proton, and no neutrons. Helium has 2 electrons, 2 protons, and 2 neutrons. Copper, which is a very common electrical conductor, has 29 electrons, 29 protons, and 34 neutrons.

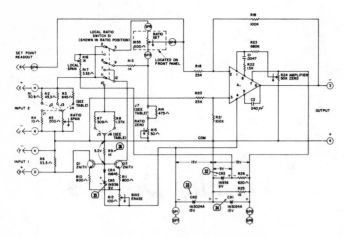

Courtesy The Foxboro Co.

Fig. 1-1. An example of an electronic schematic diagram.

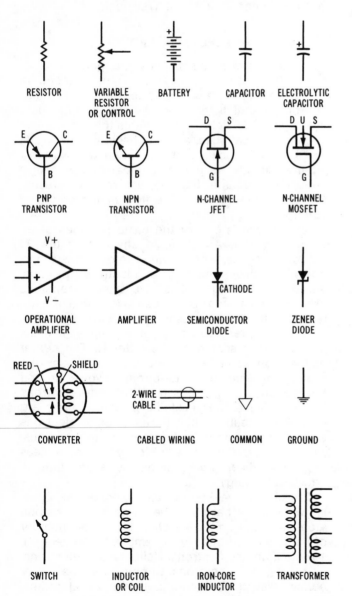

Fig. 1-2. Some common electronic symbols.

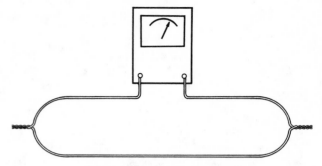

Fig. 1-3. A simple closed circuit.

It is a natural property of metals that if two different metals are joined together, the natural equilibrium of the atomic structure is disturbed in such a way that the electrons have a tendency to flow from one of the metals into the other. If, for example, copper is joined to constantan, which is an alloy of different metals, the electrons of the constantan tend to leave it and flow into the copper. If this is to happen, electrons in the copper must move to make room for the constantan electrons. This they do by flowing through the wires connected to the thermocouple and back into the constantan wire, as shown in Fig. 1-5. In this way, a current or the movement of electrons occurs.

Current is an important electrical unit that is measured in amperes. An ampere is defined as a flow of one coulomb per second. A coulomb is a unit of electrical quantity that refers to 6 280 000-000 000 000 000 electrons. This large number may be written as 6.28×10^{18}.

In practical *electronic* circuits, one ampere of current is considered to be quite large. A milli-

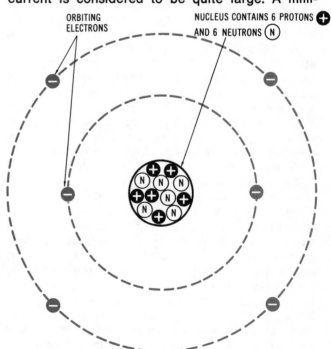

Fig. 1-4. Representative drawing of an atom of carbon.

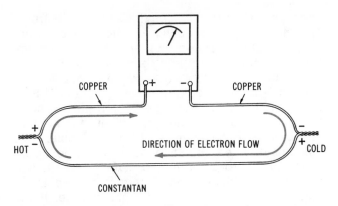

Fig. 1-5. Electron movement in a thermocouple circuit.

COPPER   COPPER

DIRECTION OF ELECTRON FLOW

HOT   COLD

CONSTANTAN

ampere (mA), which is 1/1000 (0.001) of an ampere, is a somewhat more reasonable value of current. An expression of an extremely small value of current is the microampere (μA). This value is 1/1 000 000 (0.000 001) of an ampere. Microamperes of current frequently appear in solid-state circuits and as the output of sensor circuits.

In order to produce an electron flow, an electrical force must be applied. In the case of the thermocouple just described, the dissimilar wire and the temperature difference between the two junctions of the wires will provide such a force. The force itself is called *electromotive force* and is abbreviated *emf.* The basic unit of emf is the volt.

As might be expected, all materials have a tendency to oppose the flow of electrons to some extent. This opposition, in general, is called resistance. The resistance of a material is primarily dependent upon the structure of the atoms in its physical makeup. You may recall that current is primarily based on the movement of free electrons in a material. If the material does not have an abundance of free electrons in its atomic structure, it tends to resist the flow of electrons. In practice, the resistance of a conductor is based on the atomic structure of the material, its length, cross-sectional area, and temperature. The fundamental unit of resistance is the ohm. The Greek letter omega (Ω) is used as a symbol to represent ohms of resistance. Resistance is expressed in thousands of ohms by the letter combination kΩ, and in millions of ohms by the letter combination MΩ.

Some materials are such poor conductors of electricity that they are often called insulators. Glass, porcelain, dry wood, paper, many of the plastics, and rubber are examples of common electrical insulating materials. Essentially, insulating materials do not contain an abundance of free electrons in their atomic structure.

In general, metals, carbon, ionized gases, or ionized molecules in a chemical solution make good electrical conductors. In practice, copper is the most common conductor of electricity. Silver is a better conductor, but is more costly and rarely used.

A fundamental relationship exists between the electrical units of resistance, current, and voltage. In this relationship, when 1 volt of emf is applied to 1 ohm of resistance, a resulting current of 1 ampere will occur. Another way of expressing the relationship of resistance, current, and voltage is to say that it takes 1 volt of emf to force 1 ampere of current through 1 ohm of resistance. This basic relationship is known as *Ohm's law,* which can be expressed mathematically as

$$I = \frac{V}{R}$$

where,
    I is the symbol for current,
    V is the symbol for voltage,
    R is the symbol for resistance.

Ohm's law can also be expressed mathematically as

$$R = \frac{V}{I}$$

and

$$V = I \times R$$

Notice that this law says (among other things) that current times resistance is equivalent to voltage. It says that if a current exists, a voltage also exists because all materials have some electrical resistance. In subsequent work we will frequently describe circuits where a desired voltage is obtained by causing a particular current to pass through a selected resistance. In this way it is possible to "generate" a particular voltage.

Voltage, resistance, and current are easily measured with ordinary test equipment. Voltage is often considered as a measurement of the potential difference that exists between two points. It is simply not possible to have a voltage at a single test point. It is incorrect to speak of the "voltage in a wire" or "the voltage at a terminal." When these expressions are used, what is being said is that the "voltage of a wire compared to the voltage of a second wire is of a certain value." Or, perhaps, "the voltage of a wire compared to the voltage of earth's ground is a particular value." Similarly, the expression "the voltage at a terminal" is short for saying "the voltage between terminals A and B" is a particular value.

The measurement of current, on the other hand, is a measurement of a condition at one point. Or, more exactly, it is a measurement of the number of electrons flowing past a particular point of a conductor. To make a current measurement, it is necessary to open the circuit and to cause the current to pass through the measuring instrument. In practice, this is called connecting the meter in series with the circuit under test. Notice that the current measurement of Fig. 1-5 can be made at any place in the circuit since the current is the same throughout the circuit.

The measurement of resistance is accomplished by causing a current (originating in the measuring instrument) to pass through the device in whose resistance we are interested. To make a resistance measurement, the device whose resistance is to be measured must be removed from its associated circuitry.

ELECTROCHEMICAL ACTION—
THE STANDARD CELL

There are many varieties of electrochemical cells, but their operating principles are the same. The essential components of a cell are two electrodes made of dissimilar materials connected by an electrolyte. An electrolyte is a solution that readily enables electrons to "skip" from molecule to molecule.

The electrodes are usually metals or some material such as carbon that has metallic characteristics. A common electrode combination is zinc and carbon, using sulfuric acid solution as the electrolyte. A variation of this combination is found in the so-called dry cell, which is not truly dry, since the fluid electrolyte, sal ammoniac (NH_4Cl), is made into water paste. The electrodes are the zinc outer casing and a carbon rod in the middle. The paste mixture is packed around the central carbon electrode, and the assembly is sealed with tar.

A third combination is that found in the so-called mercury cell—a "dry" type. One electrode is zinc (−) and the other is mercuric oxide (+). The electrolyte is a solution of potassium hydroxide.

The lead-acid cell, used in the common automobile battery, is composed of a lead electrode (−) and a lead oxide, PbO_2 (+) electrode, with sulfuric acid solution as the electrolyte. Some additional combinations are nickel and cadmium with potassium hydroxide, and nickel-oxide (+) and iron with an electrolyte of potassium hydroxide. The latter is sometimes called the Edison cell.

Some cells can be recharged; that is, electrical power can be put back into them. Others cannot be recharged.

The *standard* cell (or *Weston* cell) is unusual in some of its properties because, unlike some cells, if it is used properly, the voltage across its electrodes remains essentially constant. In fact, it remains so constant that the voltage can be used as a standard for comparison with other voltages; hence, the reason for calling this special cell a *standard* cell.

The Weston standard cell uses mercury as one electrode and a cadmium amalgam (mixture of cadmium and mercury) for the second electrode. The electrolyte is cadmium sulfate solution. Cadmium sulfate is formed by reacting cadmium with sulfuric acid. Sulfuric acid is a compound of the elements sulfur, oxygen, and hydrogen. Using chemical notation, sulfuric acid is written H_2SO_4, where H stands for hydrogen, S for sulfur, and O for oxygen, and the numbers indicate the number of atoms of each element in the molecule. This compound, when added to water, tends to dissociate into hydrogen ions and sulfate ions. An ion is an atom with too many or too few electrons. This results in the atom having a charge. If the hydrogen ion and sulfate ion rejoin, they neutralize each other. An electron carries a negative charge; hence, the particle lacking an electron becomes positively charged. This positively charged particle makes a good "stepping stone" for migrating electrons. In fact, the positively charged particle actively attracts electrons in an effort to get balanced.

If cadmium is added to hydrogen sulfate (sulfuric acid) it will be found that the cadmium and sulfate form a stronger compound than the hydrogen and sulfate, with the result that the cadmium drives off the hydrogen, which bubbles off as a gas, to form cadmium sulfate.

Cadmium sulfate, like the hydrogen sulfate just discussed, ionizes and becomes the "stepping stones" for the migration of electrons.

The construction of a Weston standard cell is shown in Fig. 1-6. The cell is made of glass in an H configuration. A cadmium amalgam is placed in one leg, forming one electrode. Mercury is placed in the other leg, forming the second electrode. Solid cadmium will "dissolve" into mercury (this is called an amalgam). In doing so the mercury tends to reduce the attraction between the cadmium atoms. A paste of mercury sulfate and mercury is placed over the mercury. The two electrodes are covered with a solution of cadmium sulfate, the electrolyte.

When the cell operates, the cadmium ionizes and goes into the cadmium solution. As the cad-

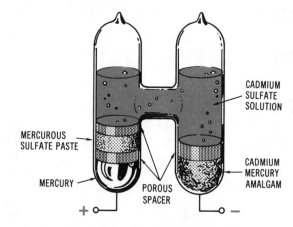

Fig. 1-6. The Weston standard cell.

mium ionizes, it liberates electrons, which pile up on the platinum lead of the mercury-cadmium electrode. The electrons are negatively charged; therefore, the lead becomes negative and is labeled with the negative sign (−). If a path is provided, the electrons will flow from the cadmium-mercury electrode to the mercury electrode. At the mercury electrode the electrons will be taken up by the mercury ions (mercurous ions) formed in the mercury electrode, deionizing the mercury and returning it to its original state. In this way a flow of electrons is caused. The pressure or electromotive force behind this flow depends on the electrode materials used. In the standard cell this emf (or voltage) is 1.0183, provided very small currents are drawn for very small periods of time. In the ordinary dry cell using zinc and carbon electrodes, the voltage is approximately 1.5. The basic voltage depends only on the electrode materials, as is the case with thermocouples. The size of the cell doesn't change the voltage—only the current-delivering capacity. If several cells are joined in series to form a battery, the voltage of the battery is the sum of the individual cell voltages.

Two methods for generating current have been considered. The thermocouple is an example of the solid-state method, wherein thermal energy (heat) is converted to electricity. The electrochemical cell is an example of converting chemical energy into electrical energy. There are other methods of generating current, particularly the mechanical to electrical method. Each of these three methods is reversible; that is, electrical energy can be converted back to thermal, chemical, or mechanical energy, for example, as in the charging of an automobile battery (electrical to chemical) and the operation of the electric motor (electrical to mechanical).

THE SLIDE WIRE AND OHM'S LAW

In the foregoing discussion, two methods of generating current have been discussed. Also, current (electron flow) was considered and it was noted that various materials resist the flow of electrons according to the composition of the material. Now the properties of electrical resistance will be con-

sidered in some detail. The slide wire will be studied as an example of some of the properties, and finally the slide wire will be considered as a demonstration of Ohm's law.

Let us examine the circuit of Fig. 1-7. The cell is an ordinary No. 6 dry cell, with a voltage of 1.5 volts between the terminals. This voltage will cause electrons to flow through the wires and the resistance. The electrons will leave the negative (−) terminal and flow through the wire and back to the positive (+) terminal, establishing a current. The amount of current is determined by how much resistance there is to the flow of electrons. The factors that affect the resistance are the kind of material used to make the wire, the length of the wire, the diameter of the wire, and, depending on the materials, the temperature of the wire.

For a given material, as we might suspect, the smaller the diameter, the greater the resistance, and the longer the wire, the greater the resistance. Some materials offer relatively small resistance. Such materials are said to be good conductors. Silver is the best conductor; copper is next, followed by gold and aluminum. Carbon is a relatively poor conductor having a resistivity about 2000 times that of copper. The resistivity of iron is about five times that of copper. Nichrome, an alloy made to have a high resistivity, is about 65 times more resistant than copper. For specific examples, 630 feet (192 meters) of 12-gauge copper wire has a resistance of 1 ohm. If the wire were iron, only 110 feet (33.5 meters) would be required to obtain 1 ohm of resistance.

Almost all conductors change resistance with temperature. This property is most useful in some cases, but a great nuisance in others. For example, the varying resistance property of some conductors makes a convenient and reliable method of temperature measurement, but if a fixed resistance is required, special alloys are needed. These alloys must be designed so that a temperature change results in a very small change in resistance —preferably no change. Manganin is an alloy widely used when variations in resistance due to temperature are undesirable. Manganin is composed of 4% nickel, 84% copper, and 12% manganese. Unlike most metals and alloys, the resistance of manganese decreases slightly as the temperature increases. The resistance of copper changes about 100 times the amount that manganin does for the same temperature change, but its resistance increases with temperature. In addition to its temperature stability, manganin has a relatively high resistivity, making it possible to obtain high resistance values with comparatively small lengths of wire.

Suppose we rebuild the circuit of Fig. 1-7, as shown in Fig. 1-8. In this figure, a resistance wire has been selected so that 10 feet (3 meters) has

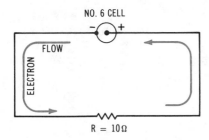

Fig. 1-7. Closed circuit with cell and resistance.

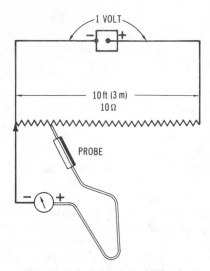

Fig. 1-8. Principle of the slide wire.

a resistance of 10 ohms. The battery has a potential difference of 1 volt between its terminals. Therefore, the wire has 1 volt applied across it. That is, between the ends of the wire, as they are attached to the battery, there is a difference in voltage equal to 1 volt. The 1 volt causes electrons to flow through the wire. Suppose we attach the negative (−) probe of a sensitive voltmeter (one whose resistance is so high that 1 volt cannot force any current through it) to the negative (−) terminal of the battery, and then touch the negative terminal of the battery with the positive (+) probe of the meter. The meter will show 0 volts. This must be so because it is impossible to obtain a voltage at one point. Now suppose we move the positive probe of the meter to the positive terminal of the battery. The meter will read 1 volt. Return the positive probe to the negative terminal and slowly draw it along the wire away from the negative terminal toward the positive terminal. The meter reading will gradually increase as the probe moves along the wire until the probe arrives at the positive terminal, where the meter will read 1 volt. Suppose we start at the negative terminal and draw the probe along the wire until the meter reads 1/2 volt and mark that point on the slide wire. If we measure the distance from that point to the negative terminal, it will measure exactly 5 feet (1.5 meters). Suppose we continue moving the probe until the meter reads 3/4 volt and measure the distance. It will be exactly equal to 3/4 of 10 feet, or 7-1/2 feet (2.3 meters). Once this was done, we would no longer need the voltmeter because we would know that at the 2-foot (0.61-meter) mark, for example, there would be a voltage difference of 0.2 volt between the negative (−) terminal and that mark; at the 6-foot (1.83-meter) mark the voltage difference between that point and the negative terminal would be 0.6 volt. Or, if we choose to compare the 6-foot (1.83-meter) mark to the positive terminal, the voltage would be 0.4 volt. Suppose we are interested in the voltage between the 2-foot (0.61-meter) mark and the 6-foot (1.83-meter) mark. This we know would be 0.4 volt.

When we marked the wire using a voltmeter, we made it possible to know what the voltages are across various points on the wire by knowing only where the marks were on the wire. Briefly stated, we calibrated the wire so that even though we measure in feet (or meters) we can speak of voltages.

Let us change the battery of Fig. 1-8 from a 1-volt battery to a 2-volt battery. The voltage drop across the wire would be 2 volts. The 5-foot (1.5-meter) mark would have a voltage of 1 volt when compared to the positive terminal. Select any other mark, say 7 feet (2.13 meters), and compare the voltage difference between it and the negative terminal. It would equal 0.7 of the applied voltage (0.7 × 2, or 1.4 volts).

Instead of using the 10-ohm resistance wire, let us replace it with a 20-ohm wire of the same length. The 2-volt battery remains the same. Again, as was the case in the previous examples, the voltage drops across the wire are related to the lengths of wire involved. The maximum voltage drop will equal the 2 volts supplied. At a point 5 feet (1.5 meters) from the negative terminal, the voltage compared to the negative terminal will equal 1 volt:

$$\frac{5}{10} \times 2 = 1$$

It is recognized that if the voltage on a given circuit is changed, something else must change. The voltage is what drives the current, and if the voltage is doubled and the circuit stays the same, the current will double. But, consider the case where the resistance of the wire is doubled and the applied voltage stays at 2 volts. Something must happen to the current. If the resistance is doubled, the current finds it twice as difficult to get through the wire; therefore, if the driving force (voltage) stays the same, it will be able to drive only half as much current through the wire.

The relationship between voltage, current, and resistance can be expressed by saying that the voltage between any two points of a circuit is equivalent to the resistance of the wire between those same two points multiplied by the current through the points. Or, in mathematical notation,

$$V = IR$$

where,

V is in volts,
I is in amperes,
R is in ohms.

Let us see what current is in the circuit of Fig. 1-7. V = 1.5 volts and R = 10 ohms; therefore, 1.5 = I ×10. Dividing both sides of the equation by 10 we get 1.5/10 = I; therefore, 0.15 = I. The current is 0.15 amperes. In Fig. 1-8, V = 1 volt and R = 10 ohms, so

$$I = \frac{1}{10} = 0.1 \text{ ampere}$$

In the example were V = 2 volts and the resistance is 10 ohms,

$$I = \frac{V}{R} = \frac{2}{10} = 0.2 \text{ ampere}$$

In the example where V = 2 volts and the resistance is 20 ohms,

$$I = \frac{2}{20} = 0.1 \text{ ampere}$$

Examine the equation V = IR and, keeping this in mind, reexamine Fig. 1-8. Notice that when a known voltage is applied to a known resistance, the equation permits us to predict how much current there will be. If we probe along this resistance, we select various resistances. When these are multiplied by the current, we are able to predict the voltages involved; and, if we go one step further and calibrate this resistance with a series of marks obtained by measuring the voltages between the marks and the negative terminal, we can predict what the voltage is by simply referring to the position on the wire.

The configuration just discussed is an example of a slide wire as used in potentiometer-type measuring instruments and in other instruments. The slide wire is a practical application of Ohm's law. In later work, the slide-wire notion will be expanded into a voltage-measuring instrument.

THERMOCOUPLE PRINCIPLES

The thermocouple is, perhaps, the only practical industrial method for measuring temperatures between 500° and 1500° Celsius. The filled system* is not designed for these high temperatures. The resistance thermometer must be specially designed if it is to be used in those ranges. For temperatures less than 500° Celsius, the thermocouple is often used, notwithstanding the fact that certain thermocouple installations cost more than a filled system would for the same job.

One of the distinct advantages of the thermocouple is that its voltage output can be readily transmitted over large distances. A second advantage is that a thermocouple can be fabricated in about 10 minutes in almost any instrument shop. The thermocouple itself is relatively inexpensive.

* See Volume 1, Chapter 5 for a discussion of filled systems.

The recording or indicating instruments used with a thermocouple may be of the null-balance type or of the deflection type. The use of null-balance instruments usually results in a higher installation cost than that for a filled system.

Emf Property of Metals

One of the natural properties of metals is that if two different metals come in contact with each other, an electromotive force (emf) is developed. The contact point is called a junction. This natural phenomenon is a property of all metals. The amount of emf developed depends on two things: the metals involved and the temperature at the junction. It is possible to take any two metals, join them, and heat the junction through a range of temperatures. A table of equivalent values can be tabulated by noting the temperature and recording the emf developed. Such data, called conversion tables, have been developed for certain combinations of metals. The emf is expressed in millivolts.

These same tables are used to convert emf back to temperature. This is the more important aspect since the purpose of the thermocouple is to measure temperature, yet the output of the thermocouple is emf. This is in some ways analogous to recording flow when the output is differential pressure.

Thermocouple Materials

Certain standard metals and pairs have been adopted for thermocouple construction. These are:

Chrome–Alumel.
Iron–Constantan.
Copper–Constantan.
Platinum–Platinum-Rhodium.

There are other combinations, but the ones listed are the most widely used. These metals are all alloys, that is, mixtures of metals. The manufacture of these alloys is carefully controlled so that a specific emf will be developed for each temperature.

The desirable properties for thermocouple materials are that the metals be relatively inexpensive and develop substantial emf for small temperature changes. In addition, the alloys should be stable over long periods of time, resistant to chemical attack, and easily manufactured to specifications. No thermocouple materials are perfect or even particularly good on all requirements. In general, the ones mentioned are adequate in the practical plant situation.

Additional Properties

In addition to the property of developing a voltage, each thermocouple pair has a definite polarity. That is, the electrons will flow (if permitted to) in a direction that depends on the materials used.

The polarities of the thermocouple materials previously listed are:

(+)	(−)
Chromel	Alumel
Iron	Constantan
Copper	Constantan
Platinum	Platinum-Rhodium

It is extremely important that polarity be observed when making connections. There is only one method of determining polarity and that is by knowing which wire is which. The wires can be identified by testing them with a permanent magnet. Alumel is slightly magnetic, iron is strongly magnetic, constantan not at all. Copper can be identified by its color.

It is important to remember that whenever and wherever two dissimilar wires come in contact, an emf is generated. It is a function of the temperature at the junction and the metals involved. The thermocouple is a utilization of this phenomenon, and certain alloys and combinations have been adopted as standard. Tables are available relating the temperature of the junction and the emf developed for the standard combinations so that it is

possible to determine the temperature if the emf is known, or to determine the emf if the temperature is known. Fig. 1-9 shows the temperature/voltage curves for the thermocouples mentioned. Observe that the curves are not straight. This means temperature charts and scales are not linear. Note that they all meet at 32° Fahrenheit (0° Celsius). The slope of the platinum–rhodium-platinum curve shows that the emf output for a given change in temperature is small when compared to iron-constantan or chromel-alumel.

THE REFERENCE JUNCTION

In the discussion on thermocouple principles, the fact was emphasized that whenever two dissimilar wires are joined a voltage will be generated. If we are to measure this voltage, it necessarily follows that an electrical circuit be completed; i.e., a closed path must be completed through the measuring instrument. In this way a continuous path is furnished for the flow of electrons. A moment's thought will show that if the circuit is closed, the two dissimilar wires are rejoined. This rejoining of the wires is accomplished in a somewhat devious manner in that the dissimilar thermocouple wires are joined through the

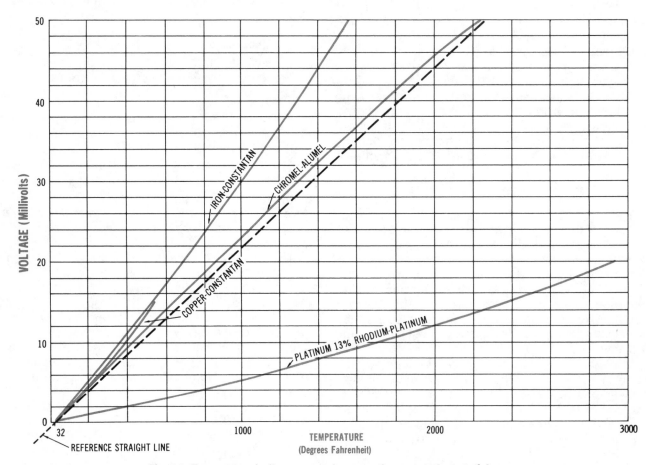

Fig. 1-9. Temperature/voltage curves for some thermocouple materials.

14

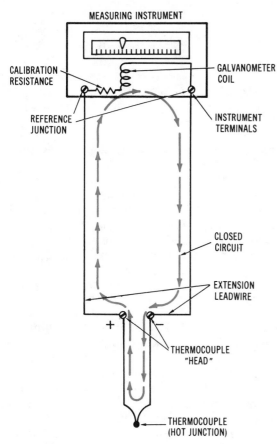

Fig. 1-10. Closed path in a thermocouple instrument.

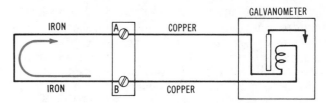

Fig. 1-11. Junctions of dissimilar metals.

measuring instrument as shown in Fig. 1-10. Nevertheless, a second junction of dissimilar wires is formed. This second junction is the reference junction or "cold" junction. The designation reference junction will be used in our discussions. Thus, it can be seen that in every thermocouple circuit there are at least two junctions of dissimilar wires. The problem then becomes one of determining how to make a temperature measurement with such a system.

Properties of the Reference Junction

It will be recalled that the voltage generated by a junction of two dissimilar metals is determined by the metals involved and the temperature of the junction. It has been stated that every circuit necessarily has two junctions. This is so because if two different wires are joined at one end, the other ends must also be joined to form a circuit. Unfortunately, the second junction (reference junction) in many cases is not physically obvious and may appear in different locations in different circuits. To further complicate the matter there are several junctions of dissimilar metals which do not form reference junctions. Consider the circuit in Fig. 1-11.

Terminal A is a junction of iron to copper, and terminal B is also a junction of iron to copper.

If terminals A and B are at the same temperature, the voltage generated at terminal A will be equal to the voltage generated at terminal B. Notice that if we proceed around the circuit in the direction indicated by the arrow, the voltage at (say) terminal B increases by 1 millivolt, but the voltage at terminal A decreases by 1 millivolt. In other words, they oppose each other, and since they are equal, they cancel each other out. As far as the contribution of terminals A and B to the voltage in the circuit is concerned, they could just as well be directly joined together.

Now let us add a thermocouple to the circuit, as in Fig. 1-12. Starting at terminal D, and proceeding in the direction shown by the arrow, observe that at terminal D, iron is joined to iron; hence, no voltage is generated. At point X (the thermocouple), iron is joined to constantan and a voltage is generated. Proceeding to terminal C, constantan is joined to iron, and at this terminal a voltage is generated. In our previous discussion it has been shown that, as far as voltages are concerned, terminals A and B could be joined. If this is done, iron is joined to iron; therefore, terminals A and B may be disregarded, leaving only point X (the thermocouple) and terminal C, both of which are junctions of iron and constantan. Notice, however, that at the thermocouple, iron is joined to constantan (when proceeding in the direction of the arrow) and at terminal C, constantan is joined to iron. This is the reference junction. The consequence is that the voltages at the two junctions oppose each other. If the temperatures at these two points are the same, the voltages at the two junctions will cancel each other. The only time that there will be a net gain in voltage is whenever there is a difference in voltage between the thermocouple and the reference junction. It is this difference in voltage that the measuring instrument "sees." In this case, terminal C is the second or reference junction. It is absolutely necessary that the temperature of the reference junction be

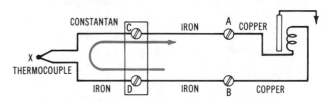

Fig. 1-12. Basic thermocouple action.

known, as well as the net voltage of the circuit, before the thermocouple temperature can be determined. The galvanometer can indicate only the millivolts of the circuit. Various means for determining the reference junction temperature are available and will be discussed later. First, however, the location of the reference junction will be considered.

Reference Junction Location

An examination of Fig. 1-12 will show two terminals, A and B, where iron and copper are joined. These have been shown to "cancel out." The actual circuits, however, are arranged somewhat differently in that the iron wire between C and A and D and B is replaced with copper wire; or a second possibility is that an iron wire may be used between D and B and a constantan wire between C and A. The overall operation is the same in either case, and is the same as the circuit of Fig. 1-12. The difference between the two methods is one of location of the reference junction. In the case where copper is used, the reference junction is located at the terminals of the thermocouple, as in Fig. 1-13. For the case of the iron and constantan wire, the reference junction is located at the instrument terminals, as in Fig. 1-14. The reference junction is always located at the junctions in the complete circuit where two different pairs of wire are found. For example, in Fig. 1-13 the reference junction is located where the pairs are constantan-copper and iron-copper. In Fig. 1-14 the same pairs are to be found at the instrument where the constantan and iron are joined with the copper of the measuring instrument. Thus, the use of constantan and iron wires resulted in the reference junction being located at the measuring instrument. The location of the junction at the instrument makes it possible to take into account the reference junction contribution to the net voltage in the thermocouple circuit.

The special iron-constantan wire is called extension lead wire. Lead wire is available in combination for use on the various kinds of thermocouples. It should be noted that the lead wire is not necessarily made up of the same material as the couples. For example, it is not uncommon to find a lead wire of iron-constantan used on a

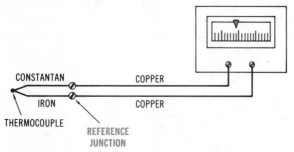

Fig. 1-13. Reference junction at thermocouple terminals.

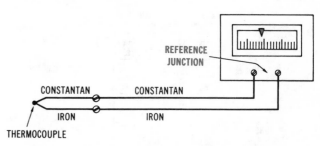

Fig. 1-14. Reference junction at instrument terminals.

chromel-alumel couple. This usage may seem to be contrary to what has been said. It is possible to mix these wires because the voltage generated by the two different combinations is substantially the same if the temperatures are less than 150° Fahrenheit (65.56° Celsius). An examination of the curves in Fig. 1-9 will show that the iron-constantan and the chromel-alumel curves coincide at a low temperature.

Reference Junction and Conversion Tables

Conversion tables are available relating the output of the several thermocouple types. It has been pointed out that the output of a thermocouple of a given material depends on the difference in temperature between the couple and the reference junction; therefore, each table is based on a known reference junction temperature. The most common reference temperatures are 0° Celsius and 32° Fahrenheit. The reason for two tables is one of convenience. The two tables make it unnecessary to convert from the Fahrenheit scale to the Celsius scale or vice versa.

If the actual temperature of the reference junction is different from the reference temperature of the table, a correction must be made.

Methods for Compensating (Correcting) for Reference Junction Temperature

It will be remembered that the net voltage in a thermocouple circuit is determined by the temperature of two junctions: the measuring junction and the reference junction. The voltage generated is determined by the difference in temperature between these two junctions. This difference is expressed in millivolts and must be converted to degrees. In other words, the signal is calibrated in terms of temperature. However, before this can be done, the reference junction temperature must be taken into consideration. The methods for doing this are called reference junction compensation.

The methods might be classified as follows: (1) paper and pencil, (2) manual, and (3) automatic. Manual and automatic compensation will be discussed with potentiometers.

The paper-and-pencil method consists of finding the millivolt equivalent of the reference junction

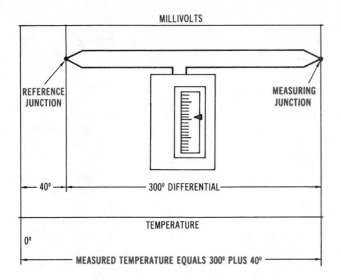

Fig. 1-15. Compensation for reference junction temperature.

temperature. In order to do this the reference junction temperature must be measured with a glass-stem thermometer. This temperature value can be located in the proper table and the millivolt equivalent can be determined. This then must be added to the differential millivolt reading of the measuring instrument. (See Fig. 1-15.) This problem is somewhat analogous to the problem of absolute pressure versus gauge pressures. However, there are further complications in that the millivolt reading has to be converted to degrees. It might be helpful to think in terms of millivolts alone for the time being, recognizing that, given a millivoltage, it can be converted to degrees.

In addition to the three methods suggested, there is a fourth possibility that deserves some comments. This method consists of holding the reference junction at a known temperature. One of the most convenient ways of doing this in the laboratory is to immerse the reference junction in a water/ice bath, as in Fig. 1-16. The temperature of this bath is 32° Fahrenheit or 0° Celsius.

To summarize, every temperature measurement made with a thermocouple is necessarily a measurement of the difference in two temperatures.

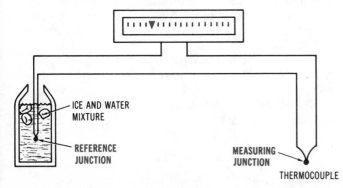

Fig. 1-16. Laboratory setup for controlling reference junction temperature.

One of these temperatures is the temperature of the reference junction. The second is the temperature of the measuring junction.

The reference junction temperature must be known before the temperature of the measuring junction can be determined. This is accomplished in different ways. The paper/pencil method and the constant temperature method have been considered; other methods will be discussed in detail in following paragraphs.

MEASURING TEMPERATURE WITH THERMOCOUPLES

Introduction

In preceding paragraphs the principle of thermocouples, the problem of the reference junction, and a method for making thermocouples have been discussed. Now we will discuss the two basic instruments used with thermocouples, and consider some of the problems in the practical use of thermocouples.

Thermocouple Measuring Instruments

The measuring instruments used with thermocouples fall into two general categories:

1. Deflectional.
2. Null.

On the surface it might appear that all measuring instruments are deflectional in that the indicator or pointer moves over the scale. However, a study of the operation of the instruments will show a clear distinction in the two types.

The deflectional type is characterized by an arrangement of the millivolt detector so that there is a change in position of the detector for each change in measured voltage. For example, consider the galvanometer mentioned previously. As the current through the coil increases, the angular rotation of the galvanometer increases. In this way the galvanometer can be used directly to measure the voltage of a thermocouple circuit.

The null method is somewhat more complicated but there are many advantages to it. The essential characteristic of a null-type measurement is that the detector assumes the same position regardless of input; that is, the position of the detector after the new input is balanced is the same as the position prior to the new input. With appropriate circuitry, this is readily accomplished.

The null and deflectional concepts will be discussed later in more detail. The various specific instruments will be classified as either null or deflectional. At this time, an example of the two different concepts can be found in weight scales. Consider the spring scale that junkmen and fishermen may use. These scales are essentially a spring fixed at one end. The object to be weighed is attached to the movable end. The greater the weight of the object the more the spring will be stretched. A pointer is fastened to the movable end of the spring. The spring movement is calibrated in pounds and ounces. This scale is a deflectional instrument.

Now consider the laboratory balance. Prior to making any measurements, the scale is balanced. That is, an indicator is brought to a midpoint on a scale. A determination of weight is made in the following manner: The unknown weight is added to one pan, deflecting the scale. Weights of known magnitude are added to the other pan until the indicator is brought back to the midpoint. The unknown weight is determined by knowing the total weight of the balancing weights added to the scale.

Notice that the scale pointer is at the same point as it was before the weight was added. This will be recognized to be the essential requirement of null measurements. The scale, as used here, provides a method for comparing the weight of an unknown body to a known weight.

Thermocouple Installations

The thermocouple in most cases is placed within a closed protecting tube that is called a thermowell. Thermowells are available in numerous sizes and materials. Their selection is based on:

1. The physical and chemical properties of the materials being measured.
2. The temperature range.
3. The dimensional requirements.

For example, extremely high temperatures may require that a ceramic well be used; corrosive acids may require stainless steel; a deep tank may call for a well 15 to 20 feet (4.6 to 6 meters) long.

Thermocouples may be welded (or soldered) in the bottom of the well. This cuts down the amount of time it takes for the temperature change to reach the thermocouple. The disadvantage of a welded (or soldered) thermocouple is that visual inspection of the thermocouple is impossible. Removal or replacement of the well cannot be made unless the process conditions permit the removal of the well. There is no practical way to check the thermocouple with a second thermocouple. In general, it would be wise to avoid thermocouples that are welded or soldered in their wells. However, where fast response is absolutely necessary, a thermocouple that is welded into the well must be used.

Lead wire is run from the junction block of the thermocouple to the recorder or indicators. If possible, the lead wire run should be a single piece of wire without any splices. If splices are necessary, great care should be exercised in making them. Mechanical connectors are not recommended. A satisfactory method is to separate the wires for several inches, clean away the insulation for approximately 2-1/2 inches (6.5 centimeters), then clean the wires thoroughly. Twist the wires to form a pigtail about 2 inches (5 centimeters) long. Fold the pigtail in halves and solder the connection. Tape the joint with plastic tape. Use great care to ensure a vapor-tight insulation. In general, four to five times the amount of tape that might be used on an electrical connection is required on a thermocouple connection. Even the best made splices are potential trouble spots and are not satisfactory in conduit that becomes moisture soaked.

The reason for the elaborate precaution with thermocouple wiring arises out of the fact that the system carries milliamperes and millivolts. The slightest leakage between the wire and the conduit may "leak out" significant amounts of current, causing errors. If the leak is from wire to wire, the effect is that of a junction which tends to short out the thermocouple.

The installation of thermocouple wiring is a much more exacting job than the installation of electrical wiring. Techniques that are satisfactory on electrical wiring may cause all sorts of problems on thermocouple wiring.

Servicing Thermocouple Systems

A breakdown of typical service calls on thermocouple systems follows:

Thermocouple failure (due to wetness, damp corrosion, age, damaged insulators, broken junction blocks, etc.) would represent about	80%
Extension lead wire failure (caused by faulty splices, polarity errors, wet or damaged insulation, especially at the thermocouple)	10%
Recording or indicating instruments	5%
Others	5%
	100%

The type of service depends to some extent on whether a null- or a deflection-type instrument is being used. In general, the null instrument is more tolerant of poor (high resistance) connections, whereas the deflectional type is more tolerant of wet or damp systems.

A thermocouple system in good repair will be satisfactory on either instrument, subject to the requirement of the deflectional instrument that the resistance of the lead wire must not exceed a certain fixed amount.

GALVANOMETERS

An important instrument component is the galvanometer. Galvanometers are the basic components of many power and current meters. Also, they are widely used in conjunction with the thermocouple to measure temperature. In our work we shall discuss two galvanometer types. The types are determined by the method of supporting the galvanometer coil. In the first type the coil is supported by pivots running in a "bearing." This type is based on the D'Arsonval meter movement. The second type is suspended by straps that twist as the coil rotates.

Galvanometers may also be classified according to usage. One classification is composed of the deflectional galvanometer, the second is the galvanometer as a null detector. However, the galvanometers themselves are the same in principle and operation. The difference comes about in the circuitry associated with the specific galvanometer. In general, the suspension galvanometer is used as a null detector, and the pivot-bearing (D'Arsonval-type) galvanometer is used as a deflectional galvanometer. There are many galvanometer varieties in addition to those mentioned. For the most part, our discussion will be limited to the D'Arson-

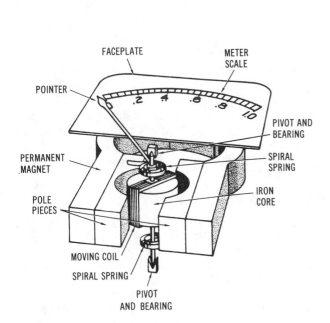

(A) Basic components of meter.

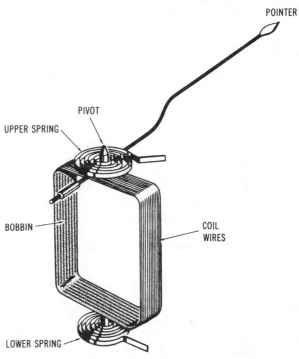

(B) Details of moving-coil assembly.

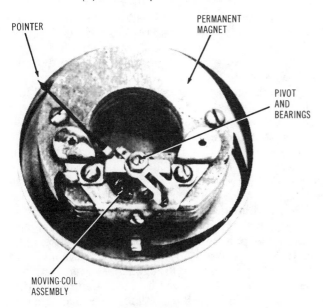

(C) Photograph of movement.

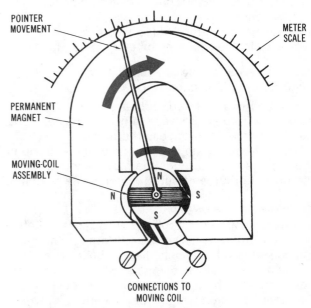

(D) Principle of operation.

Fig. 1-17. The D'Arsonval-type galvanometer.

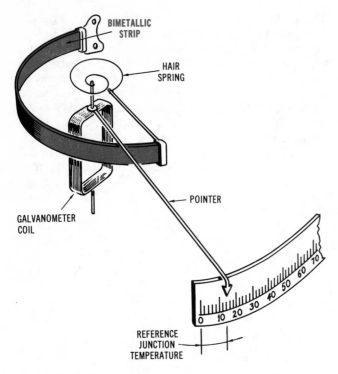

Fig. 1-18. Reference junction compensation with bimetallic strip.

val-type, direct-current galvanometer. Details of the D'Arsonval meter movement are shown in Fig. 1-17.

Components

A D'Arsonval-type galvanometer consists basically of:
1. A movable coil and pointer.
2. A permanent magnet with "pole pieces."
3. Spiral balancing springs called "hair springs."
4. A scale.
5. An RJ indicator (added when the galvanometer is used as a pyrometer). (See Fig. 1-18.)

Arrangements of the D'Arsonval-Type, Pivot-Bearing Galvanometer

The coil is supported on bearings and mounted so that it rotates between the pole pieces. The hair springs tend to oppose the rotation.

The electrical path is from one terminal through the upper spring, through the moving coil, through the lower spring, and out to the opposite terminal.

Principle of Operation

When a voltage is applied to the galvanometer, a current will result. The current through the moving coil causes a magnetic field to build up around the coil. This coil field reacts with the magnetic field produced by the permanent magnet. The coil field tends to align with the permanent field, which causes the coil to rotate. This rotation is resisted by the hair springs.

The greater the current through the coil (which means the greater the applied voltage), the stronger the coil field, and the greater the attraction between the coil field and the magnet field. The greater the attraction, the larger the force available to rotate the coil.

This larger rotation is opposed by a stronger hair-spring force, since the hair spring is under a larger deformation. Finally, the spring balances the force of the magnetic fields and the galvanometer pointer comes to rest.

As a consequence of this chain of events, the input voltage has been converted to a pointer rotation. If the pointer rotation is read with a scale calibrated in volts, a measurement of the input voltage is obtained.

Suspension Galvanometer

The coil is frequently suspended on thin straps. These "suspensions" do the same job as the hair springs and the pivots and bearings of the D'Arsonval meter. The resistance to twisting of suspension is the opposing force that balances the forces of the magnetic fields. Usually, a suspended galvanometer is used in a "null-balance" type of measurement.

Service and Calibration

A galvanometer is an extremely low-power device, and it is imperative that it be as free of friction as is possible. Therefore, the bearings and pivots must be in top condition and free from all traces of dirt and oil. The clearance between pole pieces and coil must be free of any dust, dirt, or lint. To test for cleanliness use the "tap" test. With a constant emf applied to the galvanometer, watch the pointer closely and gently tap the galvanometer. If the pointer shifts position there is a "tap error." Such an error is a mechanical stoppage which might be due to lint, dirt, metallic chips, or any other foreign body lodged between the pole pieces and the coil.

Other possible causes of "tap error" are blunt pivots and cracked bearings. The pivots may be checked for sharpness by sliding them lightly across the thumbnail. If they are in good condition, they will scratch the nail. The bearings may be examined with a jeweler's eyepiece for cleanliness and cracks. The convolutions of the hair springs should not touch each other.

The pointer is a light-weight aluminum tube that slides onto the pointer base. Shellac is used to attach the pointer to the pointer base. The pivot bases are attached to the coil by shellac. Care must be exercised so that the bases are not dislodged. If they are, a small pencil-type soldering iron is used to melt the shellac. The pivot base is then set in position and the shellac is allowed to harden.

Reference Junction Compensation

A review of thermocouple theory will show that whenever a temperature measurement is made with a thermocouple, two temperatures are involved—the temperature of the reference junction and the temperature of the thermocouple. Therefore, it is necessary that some means be developed to measure the cold-junction temperature.

Suppose that the galvanometer was caused to rotate upscale as the reference-junction temperature increased. The effect of such an arrangement would be to cause the pointer to move as the temperature of the reference junction changed. In a sense, the instrument would indicate ambient temperatures. Such devices are readily available in the form of bimetallic strips.

Bimetallic strips consist of two flat strips of different metals fastened back to back. The combination is formed into a helix. Since the strip is composed of two metals, the rate of expansion of the different metals causes the helix to unwind.

The movable end of the helix is fastened to the hair spring, and its motion is passed to the galvanometer coil, causing the coil to rotate. (See Fig. 1-18.) This corrective action can be compared to the action of the compensating Bourdon element on filled temperature systems.

BASIC POTENTIOMETER

One of the few basic electrical measuring circuits is the potentiometer circuit. A very large and important segment of electrical measuring instruments involves the potentiometer. The basis for many of the temperature measurements made with a thermocouple is the potentiometer. The following discussions are on the potentiometer as used with the thermocouple. It is important, however, that it be recognized that the potentiometer is also used in measurements of pH, infrared gas analyzers, on electrical measurements such as current, voltage, or any variable that can be converted to an electrical signal; for example, speed, r/min, thickness, and density. It would not be an overstatement to say that any and all process variables may be measured using the potentiometer.

Components

The basic potentiometer circuit consists of:

1. A source of voltage.
2. A uniform resistance so constructed that it is possible to run a contact over the length of the resistance. This resistance is called a slide wire.
3. A connection to the movable contactor.

NOTE: In general electronics work, the three-terminal variable resistance is commonly

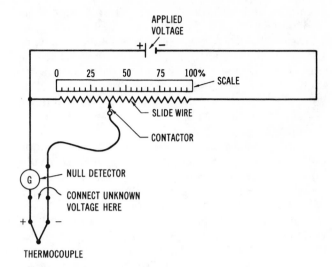

Fig. 1-19. Basic potentiometer circuit.

called a potentiometer. However, in our work we shall consider the variable resistance together with all of its associated circuitry as a potentiometer. For example, we shall describe the instrument that records temperatures measured by thermocouples as a recording potentiometer.

Arrangements

The voltage source, which is usually a battery, is connected across the slide wire. The signal to be measured is connected to one end of the slide wire and the contactor. A basic potentiometer circuit is shown in Fig. 1-19.

Principle of Operation

If the voltage appearing between the contactor and one end of the slide wire is measured, it will be found that the value of that voltage is always related to the position of the contactor on the slide wire. For example, if the contactor is at the halfway position, the voltage is equal to one-half of the applied voltage. If the contactor is at the three-fourths position, the voltage appearing at the contactor is three-fourths of the applied voltage. This relationship is true for all positions. Therefore, it is possible to determine, without measuring, what the voltage is at the contactor if the position of the contactor on the slide wire is known and if the magnitude of the voltage being applied to the slide wire is known. Thus, we have an instrument that can be used to measure an unknown voltage by comparing the unknown voltage to the known voltage generated by the potentiometer. In order to do this we require a device that will indicate when the unknown voltage is being matched by the voltage generated by the potentiometer. Such a device is called a "null detector." A galvanometer is a common null detector. Other common null detectors are electronic amplifiers and relays.

The null detector is used in the following way. One side of the null detector is connected to the unknown voltage. The other side of the null detector is connected to the zero end of the slide wire. The remaining connection of the unknown voltage goes to the contactor, thereby completing the circuit.

If there is an unbalance between the voltage appearing at the contactor and the unknown voltage, a current will pass through the null detector, causing it to indicate that an unbalance exists. It will also indicate the direction of the current. Since the direction of the current is determined by whether the unknown voltage is greater or smaller than the contactor voltage, we have a method for determining which direction to move the contactor. Suppose the unknown voltage is greater than the contactor voltage. To balance the null detector we must increase the contactor voltage. We do this by sliding the contactor to the right on the slide wire until there is no unbalance across the null detector.

The balancing action of the potentiometer can be likened to the action of a pair of balance scales as shown in Fig. 1-20. An unknown weight is placed on one pan of the scales and enough scale weights are placed on the other pan to reach a balance. The value of the unknown weight is equal to the value of scale weights required for balance.

Examine the position of the contactor as indicated by the scale. Say it is 75% of scale. Then the unknown voltage is equal to 75% of the applied voltage. Thus, we have a device that can be used to measure voltage.

It is possible to modify the basic potentiometer circuit so that the applied voltage will always be known. When this step is taken, a complete voltage-measuring device will have been obtained.

BASIC POTENTIOMETER WITH STANDARDIZATION

A limitation of the basic potentiometer circuit arises out of the fact that in order to measure an unknown voltage using the potentiometer, it is necessary to know the magnitude of the voltage that is being applied across the slide wire. The means for making known the applied voltage are referred to as the standardizing circuit or, simply, standardization.

Components

To incorporate a method for standardizing a potentiometer, the following components are required in addition to those of the basic potentiometer:

THE ACTION OF POTENTIOMETER CAN BE LIKENED TO SCALE ACTION

NULL DETECTOR

CONTACTOR VOLTAGE

UNKNOWN VOLTAGE

(ADD OR SUBTRACT VOLTAGE UNTIL BALANCED)

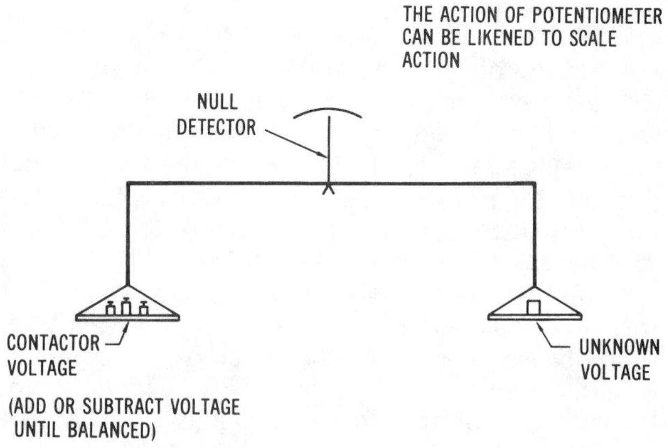

Fig. 1-20. Potentiometer action compared to balance scales.

1. An adjustable resistance called the battery rheostat.
2. A resistance called the standardizing resistance.
3. A standard cell (described earlier in the chapter).
4. A switch called the standardizing switch.

Arrangements

The battery rheostat is put in series with the battery and slide wire.

The standardizing resistance is put in series with the slide wire.

The standard cell is put in parallel with the standardizing resistance.

The standardizing switch is connected so that throwing it will connect the standard cell and disconnect the voltage under test. (See Fig. 1-21.)

Principle of Operation

The standard cell is used in a balancing action similar to that which we have just described. To use the standard cell to "standardize" the circuit,

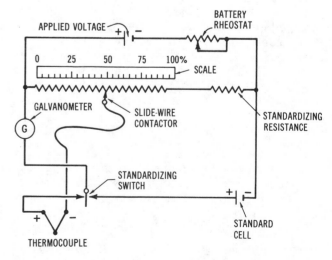

APPLIED VOLTAGE
BATTERY RHEOSTAT
0 25 50 75 100%
SCALE
GALVANOMETER
SLIDE-WIRE CONTACTOR
STANDARDIZING RESISTANCE
STANDARDIZING SWITCH
STANDARD CELL
THERMOCOUPLE

Fig. 1-21. Basic potentiometer plus standardizing components.

the standardizing switch is thrown, connecting the cell to the galvanometer and disconnecting the thermocouple. This switching applies the voltage of the standard cell across the slide wire and standardizing resistance, thereby comparing the voltage across the slide wire and resistance to the standard voltage of the cell. If these voltages are different, current will pass through the galvanometer, causing it to indicate the unbalance and also the direction of the current.

If the galvanometer is deflected, the battery rheostat is adjusted, thereby changing the current through the slide wire and resistance. This changes the voltage across the slide wire and resistance ($V = IR$). Thus, it is possible to balance the standard-cell voltage across the slide wire and resistance.

Suppose that on throwing the standardizing switch, the galvanometer indicated that a current was being delivered from the standard cell into the slide-wire circuit. This means that the standard-cell voltage is greater than the voltage across the slide wire and resistance. In order to increase the voltage across the slide wire and resistance it is necessary to cause more current to be delivered through them. This is accomplished by reducing the resistance of the battery rheostat.

This increase in current will result in an increase in voltage across the slide wire because of the relationship in which voltage equals the current times the resistance ($V = IR$). When the voltage across the slide wire and resistance equals the standard-cell voltage, the galvanometer will come to zero and the instrument is said to be standardized.

CAUTION: Remember that the standard cell has very little current-delivering capacity and, therefore, must not be connected into the circuit for long periods of time. Two or three seconds is about as long as the standardizing switch should be closed at any one time.

BASIC POTENTIOMETER WITH STANDARDIZING CIRCUIT AND REFERENCE JUNCTION COMPENSATION

It has been shown that a thermocouple is essentially a device that converts degrees of temperature to millivolts and that the equivalent millivolts are determined by the difference in temperature between the reference junction and the hot junction. This fact makes it necessary that the reference junction temperature be known or that there be a method for compensating for the reference junction temperature. The discussion will show how the reference junction is compensated for when using the potentiometer circuit to measure temperature. Two methods are discussed—manual and automatic.

Components

The components for manual compensation are a fixed resistance and a variable resistance. The variable resistance is knob operated for manual compensation. In the case of automatic compensation the variable resistance is made of special resistance wire.

Arrangements

The variable resistance is put in series with the fixed resistance. The variable and fixed resistances are put in parallel with the slide wire. The tap of the variable resistance is connected to the null detector.

Principle of Operation (Manual)

In the discussion on reference junction compensation it was shown that the voltage output of a thermocouple circuit is determined by the difference in temperature between the hot and reference junctions. Therefore, it is necessary to know the value of the reference junction temperature. This temperature is converted into millivolts and added to the millivolts output of the thermocouple circuit. This can be accomplished electrically (see Fig. 1-22) as follows:

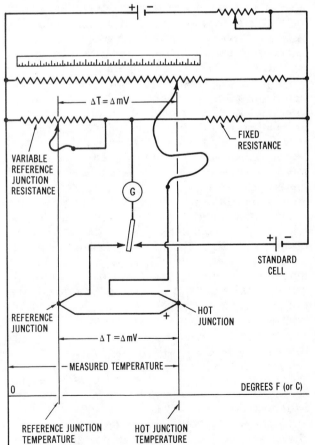

Fig. 1-22. Basic potentiometer with manual reference junction compensation.

Measure the reference junction temperature with a thermometer. Rotate the reference junction compensation, moving the contactor upscale. This upscale travel is the millivoltage equivalent of the reference junction temperature. Since one side of the null detector is connected to the movable contact, the effect of this adjustment is to move the starting point upscale; thus, a new reference point has been established and this new reference is equal to the reference junction temperature.

The thermocouple output is equal to the difference between the reference junction and the hot junction. The reference junction has been shifted upscale an amount equal to its temperature. Therefore, the hot junction can now be read directly on the scale of the instrument.

Principle of Operation (Automatic)

Suppose it were possible to get the variable resistance to change as the temperature of the reference junction changes. If this were possible, then it would not be necessary to manually move the tap, and an automatic reference junction compensation would be obtained. To get a resistance that will change as the temperature changes is not at all difficult. A resistor made of nickel wire has that property. As the temperature changes, so does its resistance. If such a resistance is used instead of the variable resistance, automatic reference junction compensation is obtained.

This reference junction compensator is located so that it will be at the temperature of the reference junction. The reference junction is at the point where the dissimilar wire of the thermocouple is rejoined, which almost invariably is at the terminal strip of the instrument. (See Fig. 1-23.)

It is interesting to note that this method of compensation results in a Wheatstone bridge arrangement. However, the instrument is still a "potential" meter, or a potentiometer. The introduction of the Wheatstone bridge method of reference junction compensation, unfortunately, does result in confusing the potentiometer with the Wheatstone bridge. All too frequently both instruments are called "bridges," but it should be remembered that the potentiometer is a voltage-measuring instrument, whereas the Wheatstone bridge is a current-measuring instrument, and in a certain sense they are opposites.

THE WHEATSTONE BRIDGE

A substantial portion of process variable measurements are made using the Wheatstone bridge. Examples are conductivity, gas analyses, temperature, and strain-gauge measurements. Fig. 1-24 shows a typical Wheatstone bridge that is often used to make industrial measurements. This particular instrument measures resistance values from

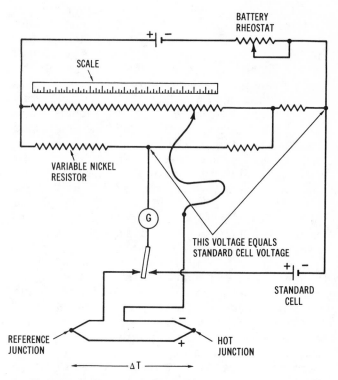

Fig. 1-23. Basic potentiometer with automatic reference junction compensation.

1 ohm to 11.01 megohms. Two resistance decades and a slide-wire resistor are built in to this instrument. All resistance values and dial indications are displayed on the central window.

The Wheatstone bridge is, in certain respects, the opposite of the potentiometer. The potentiometer has been shown to be a voltage-measuring device. The Wheatstone bridge is a resistance-measuring device. The potentiometer, since it is a voltage-measuring instrument, requires a fixed and known voltage, hence the need for the standardizing circuit. Varying resistances in the external circuit make very little difference. The Wheatstone bridge functions successfully with wide variations of applied voltage, but varying resistances (other than the one being measured) are intolerable. It is important that the basic purpose and operation of each of these two instruments be well understood.

Components

The Wheatstone bridge consists of the following components:

1. One known, fixed resistance.
2. One known, variable tapped resistance called a slide wire.
3. A null detector.
4. A voltage source.
5. The unknown resistance that completes the circuit.

Fig. 1-24. A typical Wheatstone bridge.

Courtesy Leeds & Northrup Co.

Arrangements

The unknown resistance is put in series with the fixed resistance, and this combination is put in parallel with the variable tapped resistance, as in Fig. 1-25. The tap is connected to the null detector. The other side of the null detector is connected to a point between the unknown resistance and the known fixed resistance. Power is connected to the ends of the tapped resistance.

If the bridge is to be used for temperature measurement, then the unknown resistance must change with temperature. If the bridge is to be used for measuring variables other than temperature, then the unknown resistance must be made to change as the measured variable changes. For example, if the bridge is to be used to measure the percentage of SO_2 in air then the unknown resistance must change as the percentage of SO_2 changes.

Operation

If a voltage is applied to the tapped resistance and the parallel branch composed of the unknown resistance, a current will be delivered through the tapped resistance and the parallel branch. The current will distribute itself according to the relative resistance of the two branches.

The null detector is brought to zero by moving the contactor. When the null detector is at zero, the current through it must also be zero. If the current through the null detector is zero, it means that the voltage across the null detector must be zero. The only time that this voltage is equal to zero is when the voltage drops across resistances A and C are equal. The moving of the contactor changes the resistance of C and simultaneously causes a change in the relative amount of current. This combined effect results in the current through

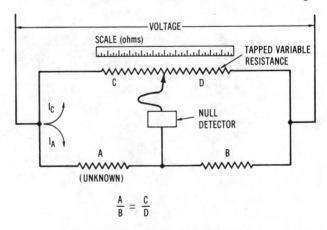

Fig. 1-25. Wheatstone bridge circuit for resistance measurement.

C times the resistance of C to equal the current through the unknown resistance times its resistance, or

$$I_C \times C = I_A \times A$$

If there is no current through the null detector, then the current through resistance C must also pass through resistance D, and the current through A must also pass through B, or

$$I_C \times D = I_A \times B$$

If the two equations are divided, the following equation is obtained:

$$\frac{C}{D} = \frac{A}{B} \text{ and } A = \frac{C}{D} \times B$$

In other words, A, the unknown resistance, equals C divided by D, times B. The values of D and B are known and the value of C is determined by the contact position on the variable resistance. If a scale is used to indicate the contact position, the value of C can be read on the scale. Hence, we have a device that measures resistance.

This instrument can be used to measure temperature if a resistor could be made so that its resistance changed with temperature. When a resistor made of platinum or nickel, for example, is enclosed in a pressure-tight bulb, it is possible to measure the temperature of certain materials by immersing the bulb in them. A Wheatstone bridge set up for temperature measurement is shown in Fig. 1-26.

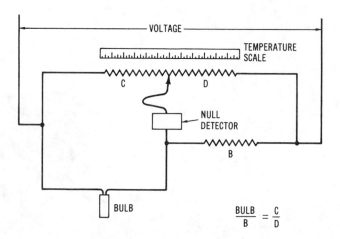

$$\frac{BULB}{B} = \frac{C}{D}$$

Fig. 1-26. Wheatstone bridge circuit for temperature measurement.

Fig. 1-27 shows a commercially available resistance thermometer bridge. This is a portable instrument with an electronic null detector that provides sufficient sensitivity to detect resistance changes of 0.0005 ohm, or values within 0.001% accuracy. Readouts are made on the central readout window.

Lead-Wire Compensation

Usually, the point of measurement is at some distance from the place where the temperature is read, making a fairly long run of connecting wire between the bulb and the instrument necessary. This wire has resistance and, what is worse, the

Fig. 1-27. A commercial resistance thermometer bridge.

resistance changes a slight amount as the temperature of the connecting wire changes. This results in an extraneous change in resistance in addition to the change in the resistance of the bulb. This action is similar to that of a filled system bulb and its connecting capillary. The lead-wire resistance change must be compensated for by running three wires to the bulb, as shown in Fig. 1-28. One wire is on one side of the null detector and a second wire is on the other side. These two wires will change resistance in equal amounts and since this change is equal on both sides of the bridge they cancel themselves out. The third wire is connected to the null detector and, at balance, there is no current through it; hence, its change in resistance makes no difference.

Properties of the Wheatstone Bridge

Since, at balance, there is no current through the null detector, the resistance between the contactor and the slide wire does not affect the accuracy of the instrument. On the other hand, any variable resistance in the connections to the instrument or in the external lead wire and the bulb connections will result in very serious errors. As a consequence, as far as it is possible to do so, all connections should be soldered. All other connections must be clean and tight.

It was shown that the absolute amount of current through each branch of the bridge was not too important, the distribution of current was the important consideration. The absolute amount of current is determined by the voltage supplied to the bridge. If this voltage changes, it does not affect the distribution of current; therefore, the accuracy of the supply voltage is not important. The only effect of a low supply voltage is that it reduces the sensitivity of the instrument.

SUMMARY

In the study of electronics, symbols are commonly used in diagrams to indicate electronic components. A few of these symbols are quite important because they are duplicated many times in schematic diagrams.

An electron is one of the basic particles of an atom. The flow or movement of electrons through an electrical conductor constitutes an electric current. One coulomb (6.28×10^{18} electrons) passing through a given point in an electrical conductor in one second is representative of one ampere of current.

An opposition to electron flow is called resistance. Resistance is measured in ohms and exists in all conductors to some extent.

Voltage is representative of the electromotive force (emf) required to drive electrons through a

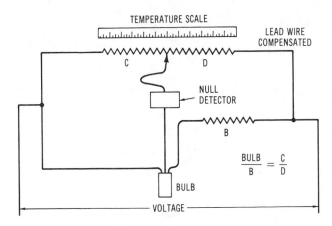

Fig. 1-28. Bridge circuit with lead-wire compensation.

conductor to overcome resistance. A voltage drop occurs when electrons flow through a resistance.

Ohm's law shows the relationship between voltage, current, and resistance. It takes 1 volt of electromotive force to drive 1 ampere of current through 1 ohm of resistance.

Thermocouples generate electricity as the result of heat applied to the junction of two dissimilar metals. These devices are used to determine temperature and serve as reference junctions in measuring applications. Thermocouple measuring applications usually employ a reference junction and a measuring junction. The measured temperature is actually a comparison in the temperature difference between these two junctions.

Galvanometers are instruments that produce a pointer deflection on a meter scale as an indication of electric current. This type of instrument employs a moving coil that is placed in a permanent magnetic field and energized by an electric current. An interaction between the coil field and the permanent magnet field causes the coil to move accordingly. A pointer attached to the coil is used to indicate values on a graduated scale.

A potentiometer is a basic electrical circuit that can be used to measure electrical voltage values. Voltage appearing across a slide-wire resistor is always proportional to the position of the contactor. Potentiometers are used by industry to measure many process variables.

Wheatstone bridges are also commonly employed by industry to measure process variables. This type of instrument is essentially a resistance measuring instrument. When a known voltage is applied to a bridge circuit, an unknown resistance value placed in one leg of the bridge is determined by nulling at the common connection point. Circuits of this type are used to determine unknown resistances that vary in value due to changes in temperature, conductivity, pH levels, etc.

CHAPTER 2

Power Supplies

INTRODUCTION

In Chapter 1, we discussed some fundamental points that were necessary to make an electronic measuring instrument operate. You should now be able to see how basic components fit together to form a simple circuit, the influence that voltage, current, and resistance have on the circuit, electrical generation, measuring devices, and potentiometer operation.

The continuation of our investigation of electronic instruments calls for the combining of components to perform specific functions. A function, for example, might change the shape of a waveform, amplify a weak signal, or change the nature of an applied voltage from ac to dc. Electronic instruments have a variety of specific functions that must be performed in order to achieve a particular operation. These functions are basic to all instruments with only a few minor component modifications. An understanding of these basic functions is an essential part of the operational theory of an electronic instrument.

In this chapter, attention will be directed toward the power supply as a functional block of an electronic instrument. This part of the instrument provides a primary source of electrical energy that makes the system operate. In its simplest form the power supply could be a battery, or combination of one or more electrochemical cells. As a general rule, only portable instruments derive their operating power from this type of source. Other instruments utilize ac from the electrical power line as a source of electrical energy.

ALTERNATING CURRENT

In industry today the primary source of electric power to operate electronic instruments is derived from the electrical power line or from portable electrical generators. As a rule, this power is in the form of alternating current or ac. As you will see in this chapter and those that follow, elec-

tronic devices such as transistors, integrated circuits, and vacuum tubes generally require dc electrical power for their operation. This means that before most electronic devices can be operated, ac power must be changed or converted into dc power. This process is commonly called rectification.

In order to understand rectification, we must first have some understanding of the nature of alternating current. In Chapter 1, we discussed the nature of direct current. You may recall that dc resulted from a constant electromotive force or voltage being applied to a closed circuit. This force caused electrons to move through the conducting material in a single direction.

Alternating current differs from direct current due to the direction of electron flow. In ac, electrons flow first in one direction for a short time, then reverse direction and flow again in the opposite direction for a short time. The flow of electrons in one direction and then in the other is called a cycle of ac. The number of cycles that occur in one second of time is called the frequency of ac. Frequency is expressed in hertz (Hz), which means "cycles per second." In the United States, the standard power line frequency is 60 Hz.

A clearer understanding of alternating current can be obtained by investigating one method by which an alternating current is generated. When a conductor is passed through a magnetic field, electrons are caused to flow in that conductor. Notice the three conditions that must be met in order to induce a current electromagnetically: the presence of a magnetic field, a conducting material, and relative motion between the two. If the poles of a permanent magnet are placed in the position shown in Fig. 2-1, the magnetic field is directed from the north magnetic pole toward the south magnetic pole. When a conductor is moved downward through the field, a current is induced into the conductor away from the observer. The symbol $\otimes$ indicates a current away from the observer, while the symbol ⊙ indicates a current toward the observer.

The preceding discussion is summed up by Lenz's law, which states: "When a conductor cuts through magnetic lines of force, a current is induced into the conductor so as to oppose the motion." This is an experimental law and can only be stated here as demonstrated by experiment. The law states that a current is induced in such a direction as to oppose the motion of the conductor. This opposition occurs because of the magnetic field produced by the current in the conductor itself. Since the electrons are flowing in the conductor away from the observer (Fig. 2-1), magnetic lines of force are produced about the conductor in a counterclockwise direction. It can be seen in the illustration that this means the lines of force below the conductor are in the same direction as the magnetic field of the permanent magnet, while the lines of force above the conductor are in the opposite direction. The effect of this is to strengthen the total magnetic field below the conductor while weakening the total magnetic field above the conductor. Thus, the current induced in the conductor produces a magnetic field about the conductor which interacts with the field of the permanent magnet in such a way as to produce a force opposite to the original force that induced the current.

The direction of the induced current described by Lenz's law can be easily remembered by a device called the left-hand rule, shown in Fig. 2-2. If the thumb and the first two fingers of the left hand are placed at right angles to one another, the first finger should point in the direction of the magnetic field, the thumb in the direction of movement of the conductor, and the second finger in the direction of the induced electron flow. In the illustration of Fig. 2-1, the direction of electron flow away from the observer is indicated by the left-hand rule.

The method of producing an alternating current that will be discussed here involves an ac generator, often called an alternator. The alternator is composed of a rotor and a stator, which is an integral part of the frame of the alternator. We shall consider an alternator in which the magnetic field is produced by the stator. This means that the stator forms the pole pieces that make up the magnetic field through which the rotor moves. These pole pieces may be either permanent magnets or electromagnets. Electromagnets utilize the principles just described, which state that magnetic lines of force are produced about a conductor through which electrons are flowing. This magnetic field is strengthened by winding the conductor into a coil around a soft-iron core. In the alternator described here, permanent magnets produce the field. The rotor or "armature" is turned or rotated through the stationary magnetic field. The armature is made up of many turns of copper wire wound about a shaft. The armature is then driven by a turbine, a gasoline engine, or some other prime mover. Electrical contact is made with the rotor through brushes that connect to the external circuit.

In the discussion of the ac generator outlined here we shall be concerned with a cross section of just one winding of the conductor. It should be remembered that what is described here takes place in many windings of the armature. The basic points to remember in this discussion are the following: Lenz's law states the direction of the induced current. The conductor must cut magnetic

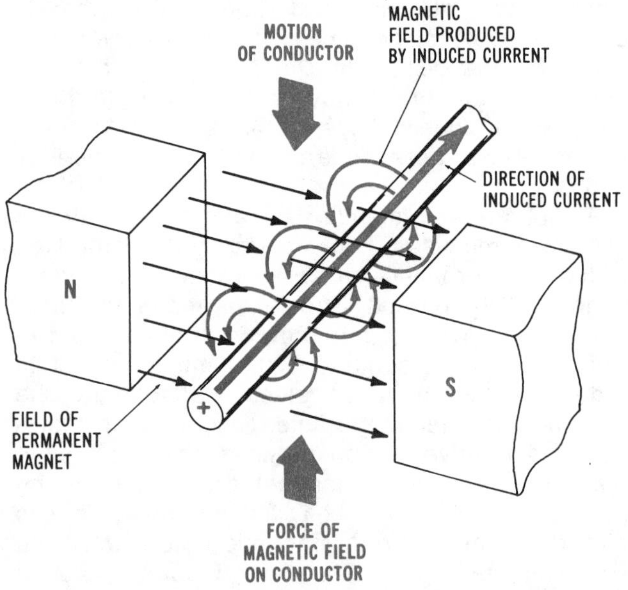

Fig. 2-1. When a conductor is moved through a magnetic field, a current is induced in the conductor in such a direction as to oppose the motion of the conductor.

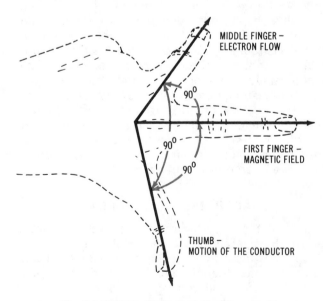

Fig. 2-2. Left-hand rule for induced current.

lines of force to induce current. No current is induced if the conductor moves parallel to the lines of force.

Now let us consider the situation in Fig. 2-3 where the conductor is rotated through the magnetic field. In position A, the conductor moves parallel to the magnetic lines of force; hence, no current is induced into the conductor, as indicated in the graph at point A. To get to position B, the conductor must move downward through the field. When this occurs, current is induced into the conductor. Remember that the direction of induced current under these conditions is away from the observer. We will call this the positive direction as indicated on the diagram. Eventually the conductor is again moving parallel to the magnetic field in position C. The electron flow at this position has again decreased to zero. Moving toward position D, the conductor is moving upward through the magnetic field and is again cutting through magnetic lines of force. Lenz's law states that induced current is now moving toward the observer. This you will note is opposite to the direction of the current observed at point B. This direction is labeled negative in the diagram. As the conductor continues to rotate in the clockwise direction, it returns to point A and the current again drops to zero. The alternator then repeats this series of operations with each rotation of the rotor.

The amount of current induced into the conductor at any given time is determined by the number of lines of force cut per unit of time. The points of maximum induced current, therefore, occur at points B and D. Between points A and B, current is produced; but since the motion is not perpendicular to the field, fewer lines of force are cut than at point B. Less induced current is the result. A gradual rise and fall of current then occurs between the zero and maximum points. The current so generated follows the pattern of the mathematical sine function and is, therefore, called a sine wave. Alternating current usually varies in this manner although many other waveforms will be encountered.

Let us investigate the characteristics of alternating current. Fig. 2-4 is a graphical representation of an ac sine wave. This graph will also represent an ac voltage, since, by Ohm's law, V depends directly on I. The maximum voltage or current occurs at point B. This is called the amplitude of this

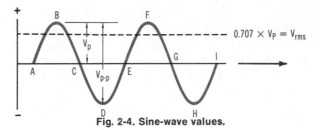

Fig. 2-4. Sine-wave values.

signal. The amplitude can also be measured at points D, F, and H, and is sometimes referred to as the peak voltage (V_p), or peak current (I_p). Also, ac signals are measured by their peak-to-peak voltage (V_{p-p}). In a sine wave, V_{p-p} is twice the value of V_p. The ac voltmeters measure the rms (root-mean-square) value of a sine wave. This is the effective value of an ac voltage or current. This means that an ac voltage with an rms value of 117 volts ac will do the same effective amount of work as 117 volts dc. The rms value of an ac voltage or current is 70.7% of the peak value. The peak voltage or amplitude of the ordinary 117-volt ac line voltage is, therefore, about 165 volts ac.

In Fig. 2-4, the portion of the sine wave between points A and C is called the positive alternation. The portion between points C and E is the negative alternation. The complete sine wave, from point A to point E, is called one cycle of the sine wave. The sine wave repeats itself between points E and I. Two cycles of an ac voltage are shown in Fig. 2-4. The time it takes for an alternating current to complete one cycle is called its period. The number of cycles that occur in one second is an indication of the frequency in hertz. Ordinary line voltage has a period of 1/60 of a second when the frequency is 60 Hz.

TRANSFORMER ACTION

Power supplies of an electronic instrument rarely operate with ac obtained directly from the power line. A transformer, for example, is commonly used to step the line voltage down or up to a desired operating value. A schematic representation of a simple transformer is shown in Fig. 2-5.

The winding on the left of the diagram is called the primary winding to which the input voltage is applied. The winding on the right is the secondary winding from which the output voltage is taken. The parallel lines between the two windings represent the core. In a power transformer, this core

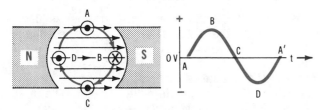

Fig. 2-3. Generation of a sine wave.

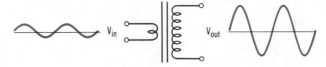

Fig. 2-5. Transformer action.

is usually made of laminated soft iron. Both windings are wound on the same core. It is not uncommon today for transformers to employ alternate primary windings to accommodate different line voltages, and to have more than one secondary winding.

The output voltage of a transformer may have the same polarity as the input voltage, or it may have the opposite polarity. This depends on the way the secondary winding is wound. A dot is sometimes used to indicate the two terminals having the same polarity. The output of the "ideal" transformer depends on the turns ratio of the transformer. If the primary winding is composed of 100 turns of wire, and the secondary has 600 turns, the transformer has a one-to-six turns ratio. Since the secondary has the most turns, it is called a step-up transformer. The output voltage will be six times the input. For example, if the input voltage in Fig. 2-5 is 3 volts ac, V_{out} equals 18 volts ac. When the secondary winding contains fewer turns than the primary, the voltage will be stepped down. Current varies inversely with the voltage. When the voltage is stepped up, the current will be stepped down. The output current also depends on the turns ratio of the transformer; i.e., if the voltage is stepped up six times, the output current will be one-sixth of the primary current. Of course, no transformer is ideal; therefore, V_{out} will be slightly less than the values described.

The question naturally arises as to how this output voltage is developed. It can be demonstrated that when a current passes through a conductor, a magnetic field is built up around it. This field is concentric with the conductor and is in a direction that opposes electron flow. The direction of this field is demonstrated in Fig. 2-6 by the use of the left-hand rule. If the thumb of the left hand is pointed in the direction of electron flow, the natural curve of the fingers follows the direction of the field. When the conductor is wound into a coil, the magnetic field of each turn of the conductor strengthens the field created by its neighboring turns of wire, forming an electromagnet.

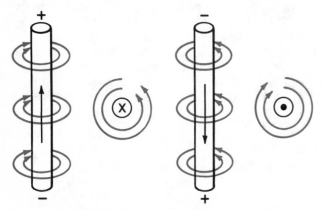

Fig. 2-6. Magnetic field around a conductor.

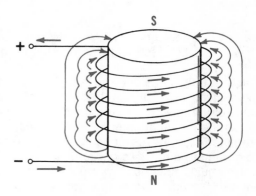

Fig. 2-7. The magnetic field of a coil.

As can be seen in Fig. 2-7, if the fingers of the left hand curve in the direction of electron flow around the coil, the thumb points in the direction of the north pole of the electromagnet. When an alternating current is applied to such a coil, the magnetic field is constantly changing. At the beginning of a cycle, the current is zero. No magnetic lines of force are present. As electrons begin to flow in the coil, magnetic lines of force begin to build up around the coil. As the current rises, the magnetic field becomes stronger. The stronger magnetic field results in more magnetic lines of force that expand and move away from the center of the coil. The magnetic field then starts collapsing as the current passes its peak. At the zero point, the field has completely disappeared. The current then begins to move in the opposite direction in the coil. The lines of force are now in the opposite direction. The end of the coil that was previously a north magnetic pole now becomes the south pole. The magnetic lines of force again expand about the coil and then collapse.

In the transformer, these expanding and collapsing lines of force cut through the conductor forming the turns in the secondary winding of the transformer. Remember that all of the conditions necessary for inducing a current into the secondary winding of the transformer are present. A voltage is then available at the terminals of the secondary winding. The more turns of wire present in the secondary the more lines of force are cut by the secondary. Therefore, a higher voltage is developed across the secondary of the transformer.

THE RECTIFIER

Assume now that a power transformer is used to step down the 117-volt ac line voltage to approximately 24 volts ac. A power supply that uses this type of transformer must then change the ac voltage into a usable form of dc voltage for the operation of a voltage amplifier circuit. The first step in changing alternating current to direct current is rectification. This process involves elimi-

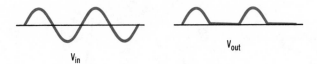

Fig. 2-8. Half-wave rectification of a sine wave.

nating one alternation of the sine-wave input so that the output only flows in one direction. Fig. 2-8 shows a graphic representation of the rectification process.

In the rectification of a sine wave, you will notice that the output only displays the positive alternation. The negative alternation, in this case, has been eliminated by the rectifier. The resulting output of the rectifier is then called pulsating dc. In practice, either the positive or negative alternation can be removed by rectification. When the positive alternation appears in the output, the developed dc is positive with respect to a common point or ground. A negative dc output with respect to ground appears when the positive alternation has been removed. Since only half of the ac wave appears in the output, this is called half-wave rectification. Both half- and full-wave rectification will be discussed in the next section.

In a power supply, rectification is normally achieved by a solid-state diode. This type of device contains two electrodes called the anode and the cathode. It has an unusual property that will let electrons flow easily in one direction but not in the other direction. As a result of ac applied to a diode, electrons only flow when the anode is positive and the cathode is negative. Reversing the polarity of voltage applied to a diode will not permit electron flow.

In order to understand the rectification process and the operation of a diode, we must briefly discuss some of the electrical properties of the elements silicon and germanium. Silicon primarily, and in some cases germanium, are the basis of nearly all solid-state electronic devices today. Fig. 2-9 shows the structure of silicon and germanium atoms. Note that the atomic number of silicon (Si) is 14 and germanium (Ge) is 32. This means that these elements have 14 and 32 orbiting electrons, respectively. Note also that they are similar to the extent that both atoms have four electrons in their outer shells.

In a crystal structure of pure silicon, such as shown in Fig. 2-10, the four electrons in the outer shell of each atom tend to join together. In this structure, which only shows the nucleus and the outer shell electrons, atoms are connected together in what is called covalent bonding. This structure forms a strong binding force that tends to hold the atoms together in a manner similar to the way that the nucleus of an atom attracts its own electrons. As a result of this structure, a crystal of silicon does not readily accept new electrons nor

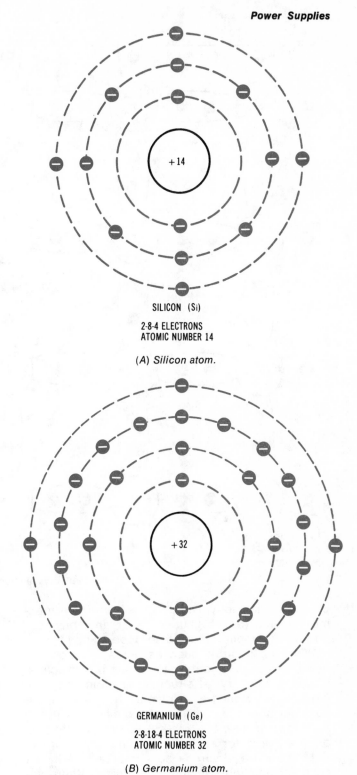

SILICON (Si)
2-8-4 ELECTRONS
ATOMIC NUMBER 14

(A) Silicon atom.

GERMANIUM (Ge)
2-8-18-4 ELECTRONS
ATOMIC NUMBER 32

(B) Germanium atom.

Fig. 2-9. Structure of silicon and germanium atoms.

does it permit bonded electrons to be removed or move around. A crystal of pure silicon, therefore, has rather poor conduction characteristics.

When a pure form of silicon or germanium is formed, it makes a good insulating material. If a minute amount of certain impurities is added to the silicon crystal, its conduction characteristics

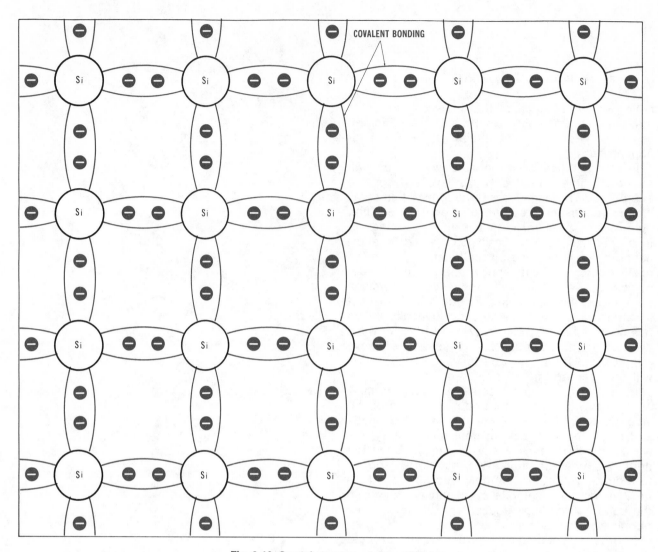

COVALENT BONDING

Fig. 2-10. Crystal structure of pure silicon.

change significantly. This process is known as doping. Arsenic (As), bismuth (Bi), or antimony (Sb) are typical impurities used to dope silicon.

When an impurity such as arsenic, which has five outer shell electrons in its atomic structure, is added to a crystal structure of atoms containing four outer shell electrons, such as silicon, the covalent bonding structure of the silicon is upset. As shown in Fig. 2-11, the structure now has an extra electron for each atom of the doping material. The extra electrons do not take part in the covalent bonding structure. As a result, there are extra electrons in the structure that are free to move around and contribute to the conduction of current. A doped crystal of silicon, therefore, becomes a conductor when voltage is applied to it. This type of structure is commonly called an *n-type* material because it has extra electrons, or negative charges, that do not take part in the covalent bonding arrangement. An n-type material will conduct current very effectively in either direction.

In the same manner, a *p-type* material is formed by doping silicon with a material that has only three outer shell electrons, such as gallium (Ga), boron (B), or indium (In). In this structure, as shown in Fig. 2-12, electron voids, or "holes," appear in the covalent bonding structure. A hole is often described as an electron deficiency spot, which represents a positively charged area. A p-type material has an abundance of positively charged areas, or electron "holes," in its covalent bonding structure and, therefore, becomes a good electrical conductor because it permits the "movement" of holes. Hole flow is in a direction opposite to that of electron flow. A p-type material will also conduct current in either direction very effectively.

When n-type and p-type materials are fused together, a solid-state diode is produced. As noted in Fig. 2-13, the p-type material represents the anode element and the n-type material serves as the cathode.

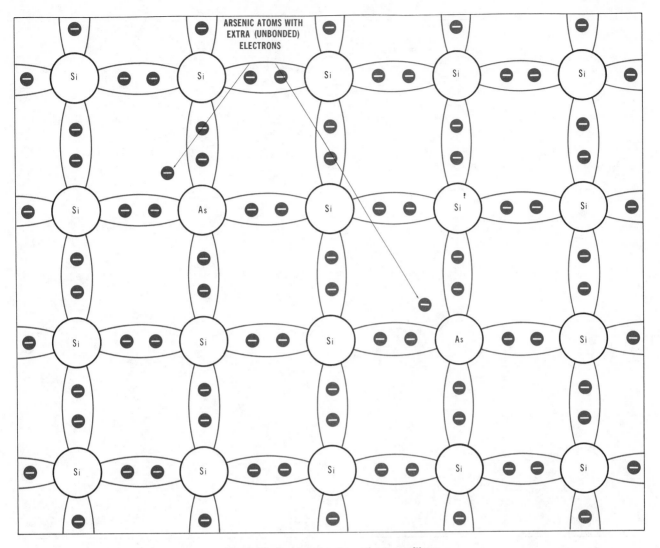

Fig. 2-11. Crystal structure of n-type silicon.

A silicon diode has the property of conducting current when connected as indicated in Fig. 2-14. In this condition, the negative side of the dc source forces electrons into the n-type material and drives them to the center or junction area. Holes, or electron voids, in the p-type material are driven to the junction area because the p-type material is connected to the positive side of the dc source.

When holes and electrons meet at the junction of a silicon diode, they combine and permit conduction of current. The process is continuous. For each hole filled with an electron, a new hole is formed by removing an electron from the p-type material. Electrons are also supplied to the n-type material as a result of its connection to the negative side of the dc source. The resistor in the circuit (R_L) is used to limit the current to a reasonable value.

When a silicon diode is connected so that the anode, or p-type material, is positive and the cath-ode, or n-type material, is negative, it is called forward biasing. Forward biasing causes a pn junction to become rather low in resistance, which in turn permits a substantial amount of current to be conducted through the junction.

When the voltage of a dc source is connected as indicated in Fig. 2-15, the diode is reverse biased. In this condition, electrons of the n-type material are attracted to the positive side of the dc source, while holes are attracted to the negative side of the source. As a result of this, the electrons and holes are drawn away from the junction. They cannot effectively combine and produce any significant amount of current when connected in this manner. Reverse biasing, therefore, forms a very high-resistance junction that permits no significant conduction of current.

When alternating current is applied to a silicon diode, it forward biases the pn junction during one alternation, and reverse biases it during the other alternation. As a result, the diode conducts heavily

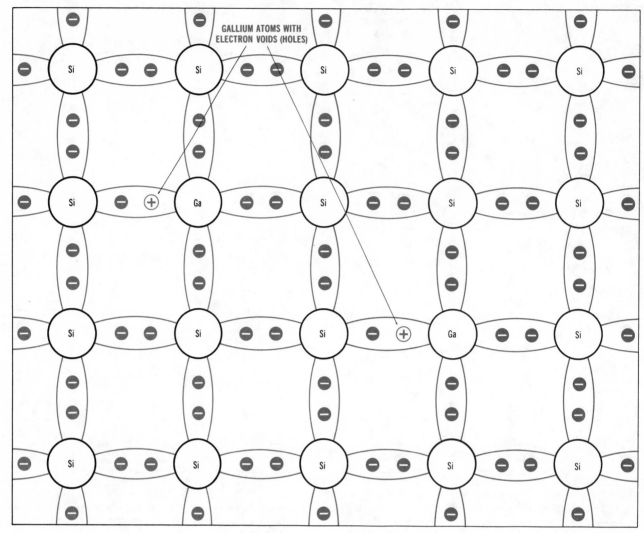

GALLIUM ATOMS WITH ELECTRON VOIDS (HOLES)

Fig. 2-12. Crystal structure of p-type silicon.

in one direction and is nonconductive in the other direction. Alternating current is, therefore, changed into pulsating direct current by the action of the diode. The terms diode and rectifier are frequently used interchangeably in electronic discussions.

HALF-WAVE AND FULL-WAVE RECTIFICATION

Rectified power supplies today are either of the half-wave or full-wave type depending on the needs of the circuit to which they are connected. Fig. 2-16 graphically shows the output differences in half-wave and full-wave rectification. Note particularly the ripple frequency of the outputs and the potential output voltage values that can be developed.

Half-Wave Rectification

Fig. 2-17 shows a simplified schematic diagram of a half-wave rectifier. In this circuit, the primary input voltage is 117 volts rms. Through normal transformer action the primary voltage is stepped down; in this case, to 25 volts rms. The positive and negative alternations of the ac voltage across the secondary winding follow in step with those of the voltage applied across the primary winding.

Assume now that the top of the transformer secondary is positive and the bottom is negative during the positive alternation of the sine-wave input. When this occurs, the diode is forward biased and permits current conduction through the load resistor, R_L, as indicated. The positive alternation will appear across the load resistor at nearly the same voltage value as that applied. The slight difference in output is due to the resistance of the diode. A silicon diode will produce a 0.7-volt drop. As a result, the output is reduced by this amount.

During the negative alternation of the sine-wave input, the top of the transformer will be negative and the bottom will be positive. By tracing the cir-

DIODE SYMBOL

ANODE ▶ CATHODE

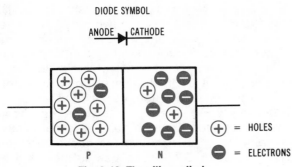

⊕ = HOLES

⊖ = ELECTRONS

Fig. 2-13. The silicon diode.

cuit back to the diode, you will notice that this voltage polarity reverse biases the diode. A reversed-biased diode represents an extremely high-resistance junction and serves as an open circuit. With no resulting current through R_L, the output voltage is zero for this alternation.

Through normal forward and reverse bias conditions, a half-wave rectifier will only produce one alternation in its output. The polarity of the output depends on the diode connections. In Fig. 2-17, the top of R_L will be positive and the bottom negative. By reversing the anode and cathode connections of the diode, the polarity of the output can be reversed. Depending on the needs of the circuit to which a power supply is connected, its output can be either positive or negative with respect to a common point or ground.

To determine the potential dc output of a half-wave rectifier, one needs to first determine the peak value of one alternation. This is achieved by multiplying the rms input voltage by 1.414. The average value of this alternation is then determined by multiplying the peak value by 0.637. Since only one alternation of the two in a cycle appear in the output, this value must be divided by 2. Combining these values we find that 1.414 times 0.637 divided by 2 equals 0.450 359, or 45% of the rms input voltage. This means that the potential dc output of a half-wave rectifier is 45%

ANODE ▶ CATHODE

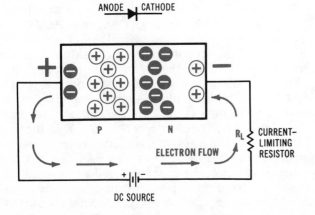

Fig. 2-14. Forward-biased diode.

ANODE ▶ CATHODE

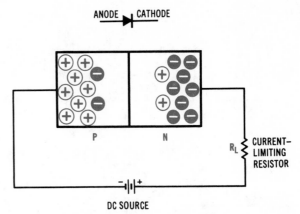

Fig. 2-15. Reverse-biased diode.

of the rms ac value minus the diode voltage drop of 0.7 volt. A dc voltmeter would read this value across the load resistor, R_L, in our simple half-wave rectifier.

The output pulses of a half-wave rectifier occur at the same rate as they do in the applied ac input. With 60-Hz input, the ripple frequency of a half-wave rectifier is 60 Hz. Only one pulse is produced, however, for a complete sine-wave input that has two alternations per cycle. Potentially, only 45% of the ac input is transformed into a usable output in half-wave rectification. Due to the low efficiency factor, half-wave rectification is not commonly used in electronic instruments that are used in industry.

Full-Wave Rectification

Full-wave rectification, as the name implies, changes both alternations of a sine-wave input into a pulsating dc output. In practice, full-wave rectification can be achieved with two diodes and a center-tapped transformer, or four diodes in a bridge circuit. Applications of both types are quite numerous today.

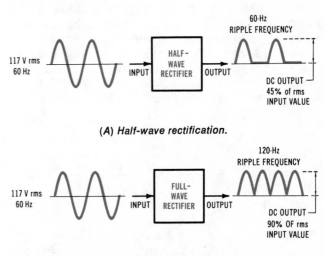

(A) Half-wave rectification.

(B) Full-wave rectification.

Fig. 2-16. Comparison of half-wave and full-wave rectification.

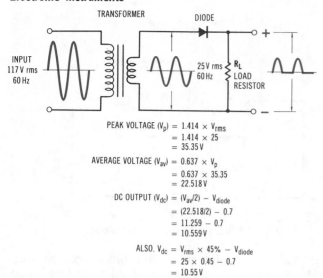

$$\text{PEAK VOLTAGE } (V_p) = 1.414 \times V_{rms}$$
$$= 1.414 \times 25$$
$$= 35.35 \text{ V}$$

$$\text{AVERAGE VOLTAGE } (V_{av}) = 0.637 \times V_p$$
$$= 0.637 \times 35.35$$
$$= 22.518 \text{ V}$$

$$\text{DC OUTPUT } (V_{dc}) = (V_{av}/2) - V_{diode}$$
$$= (22.518/2) - 0.7$$
$$= 11.259 - 0.7$$
$$= 10.559 \text{ V}$$

$$\text{ALSO: } V_{dc} = V_{rms} \times 45\% - V_{diode}$$
$$= 25 \times 0.45 - 0.7$$
$$= 10.55 \text{ V}$$

Fig. 2-17. Half-wave rectifier circuit.

A basic full-wave rectifier using two diodes is shown in Fig. 2-18. The cathodes of both diodes, as you will note, are connected commonly together to form the positive output. The anodes of each diode are connected to opposite ends of the transformer secondary winding. The load resistor (R_L) completes the circuit path between the cathodes and the transformer center-tap connection.

When ac is applied to the primary winding of the transformer, it steps the voltage down in the secondary winding of our example. If a greater output is desired, a transformer with a larger secondary voltage could be selected. Since the center tap, or point C in the diagram, is the electrical center of the transformer secondary winding, half of the induced voltage will appear between points C and A and the other half between points C and B. The resulting voltages will always be 180° out of phase with respect to point C.

As an example, assume that the positive alternation of the input causes point A to be positive and point B to be negative with respect to point C. During the next alternation, the polarity will reverse with point A being negative and point B positive with respect to point C. Point C will always be negative with respect to the positive end

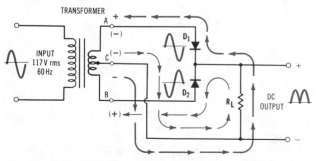

Fig. 2-18. Full-wave rectifier circuit using two diodes.

of the winding regardless of whether it is point A or point B. Conduction of the two diodes changes back and forth with one conducting during the positive alternation and the other conducting during the negative alternation. All of this is done with respect to point C. In this case, point C is considered the negative output of the rectifier.

Referring to Fig. 2-18 again, assume that the ac input is such that it causes point A to be positive and point B to be negative for one alternation. Starting at point C, electron flow is indicated by the solid arrows. With this polarity, note that D_1 is forward biased and D_2 is reversed biased. Conduction, therefore, occurs from point C, through R_L, through D_1, to point A. This produces a corresponding alternation of voltage across R_L and a resulting current through the indicated circuit path.

Using the same procedure just mentioned, assume now that the next input alternation is applied to the transformer input. In this case, point A becomes negative and point B becomes positive with respect to point C. Starting at point C, electron flow is indicated by the broken arrows. With this polarity, D_2 is forward biased and D_1 is reverse biased. Conduction now occurs from point C, through R_L, through D_2, to point B. This produces a corresponding alternation of voltage across R_L and a resulting current through the indicated circuit path.

An interesting point should be noted about the full-wave rectifier circuit. The current through R_L is in the same direction for each alternation of the input. This means that we have obtained dc output for both halves of the sine-wave input, or full-wave rectification.

The resulting output voltage of a full-wave rectifier is 90% of the ac rms voltage appearing between the center tap and the outer ends of the transformer. This value is twice the possible output of the half-wave rectifier previously discussed. This value is determined by calculating the peak value (V_p) of the rms voltage, then multiplying it by the average value (V_{av}). The potential dc output will, therefore, be the rms value times 1.414 times 0.637. Since 1.414 times 0.637 equals 0.900 718, this is a very close approximation of the 90% value. Full-wave rectification is considered to be 50% more efficient than half-wave rectification.

The dc output voltage appearing across R_L of the circuit will be slightly less than the 90% times the rms value just described. Each diode, for example, has a voltage drop of 0.7 volt. This means that a dc voltmeter would read the rms value times 90% minus 0.7 volt across R_L.

The ripple frequency of our full-wave rectifier is also different compared with the half-wave circuits. With each alternation producing an output across R_L, the ripple frequency will be twice the alternation frequency. In this case, the ripple frequency is 120 Hz or twice the input frequency. A higher ripple frequency and characteristic output of a full-wave rectifier is easier to filter than a similar half-wave output.

Full-Wave Bridge Rectifiers

A bridge structure of four diodes is commonly used in power supplies today to achieve full-wave rectification. In this rectifier configuration, two diodes will conduct during the positive alternation and two will conduct during the negative alternation. A bridge rectifier does not necessitate a center-tapped transformer as was used in the two-diode rectifier.

Fig. 2-19 shows a schematic diagram of a full-wave bridge rectifier. In this circuit, note that the ac input is applied to the junction of diodes D_1 and D_2 at the top of the bridge, and to the junction of D_3 and D_4 at the bottom. The connections of the diodes at these two junctions are reversed with respect to the other. The junction of D_1 and D_4 has both cathodes connected together, while the junction of D_2 and D_3 has the anodes connected together. The common-anode connection serves as the negative output of the bridge, with the common-cathode connection serving as the positive output.

When ac is applied to the primary winding of the power transformer, it can either be stepped down or up depending on the value of the dc needed. In our example, 25 volts ac appears across the secondary winding. In normal operation, one alternation of the input voltage will cause the top of the transformer to be positive and the bottom negative. The next alternation will cause the polarity to reverse. Opposite ends of the transformer will, therefore, always be 180° out of phase with each other.

As an example, assume that the positive alternation of the input causes point A to be positive and point B to be negative in Fig. 2-19. With this polarity, diode D_1 will be forward biased and D_2 will be reverse biased at the top junction. At the bottom junction, diode D_3 will be forward biased

and D_4 will be reverse biased. When this occurs, electrons flow from point B, through D_3, through R_L, through D_1, to point A. This path of electron flow, which is indicated by solid arrows, produces an alternation across R_L.

With the next alternation of the ac input, point A becomes negative and point B becomes positive. When this occurs,, the bottom diode junction is of positive polarity and the top junction becomes negative. In this condition, D_2 and D_4 are forward biased, while D_1 and D_3 are reverse biased. The resulting electron flow, indicated by the broken arrows, starts at point A, goes through D_2, through R_L, through D_4, to point B. This path of electron flow causes the second alternation to appear across R_L.

It is interesting to note that the current through R_L is the same direction for each alternation of the applied ac input. This, of course, means that ac is changed or rectified into dc. The dc output, in this case, has a ripple frequency of 120 Hz. Since each alternation produces a resulting output pulse, the ripple frequency is twice the value of the alternation frequency, or 2×60 Hz $= 120$ Hz.

The dc output voltage appearing across R_L of the bridge circuit will be somewhat less than 90% of the applied rms value. Each diode, for example, produces a 0.7-volt drop when conducting. This means that two diodes, that conduct during each alternation, will reduce the output by 2 times 0.7, or 1.4 volts. The resulting dc output will, therefore, be 90% of the rms value less 1.4 volts. In our circuit, this will be $0.90 \times 25 - 1.4 = 21.1$ volts dc.

Bridge rectifiers are commonly used in electronic power supplies for instruments today. A very high percentage of all solid-state industrial instruments employ this type of rectifier because of its simplified operation and desirable output. Generally, the diode bridge is housed in a single device that has two input and two output connections. A variation of the bridge circuit, using a center-tapped transformer, is also used to develop +15 and −15 volts dc with respect to a common or ground connection. Split dc power supplies of this type are needed to supply energy to operational amplifier ICs. This variation of the bridge circuit will be discussed in a later chapter.

FILTERING

After pulsating dc has been produced by a rectifier, it must be filtered in order for it to be usable in a power supply. Filtering involves changing the ripple frequency of a rectifier output into a constant dc voltage value. In practice, this is often called smoothing out the ripples of the pulsating dc voltage.

The filter circuit of Fig. 2-20 is composed of two resistors, R_1 and R_2, and three capacitors, C_1,

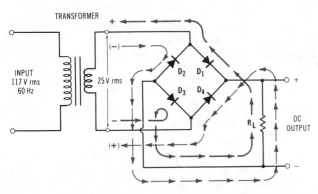

Fig. 2-19. Full-wave bridge rectifier circuit.

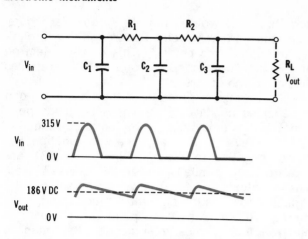

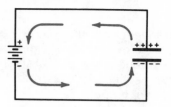

Fig. 2-21. Capacitor charged by a battery.

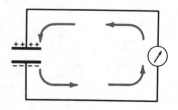

Fig. 2-20. Filter circuit.

C_2, and C_3. Resistor R_L represents the load of the power supply. This, in general, is representative of the combined electronic components receiving energy from the power supply. The filter circuit shown is called a double pi-section filter. Resistor R_1, capacitor C_1, and capacitor C_2 are connected together like the Greek letter pi (π) as are R_2, C_2, and C_3. We will first discuss a single pi-section filter, but before that we need to learn something about capacitors and their effects in dc circuits.

CAPACITANCE

Capacitance is a basic principle that must be understood in order to understand the operation of electronic circuits. The device that contributes capacitance to the circuit is a capacitor. A capacitor can be physically described as two conductors separated by an insulating material called a dielectric. Dielectric materials commonly used in capacitors are air, paper, plastic film, glass, mica, and oil. The unit for measuring capacitance is the farad. Since this is an extremely large and impractical unit, capacitors are normally measured in microfarads (millionths of farads) or picofarads (millionths of microfarads).

The capacitance of a capacitor is stated by the formula $C = KA/d$. The K in the formula stands for the dielectric constant. The dielectric constant of a vacuum is one; for air it is slightly larger than one. The dielectric constant for mica is about 7 to 9. The letter A represents the effective area of one plate of a capacitor. From the formula it can be seen that the larger this area is, the greater the capacitance will be. The letter d represents the distance between plates. It can be seen from the formula that the closer the plates are together, the greater will be the capacitance. A paper capacitor can be easily examined in order to get an idea of its construction. It is formed by using two strips of metal foil as the plates, with leads (pigtails) connected to each of these "plates." Between

the plates are inserted strips of paper that act as the dielectric. In order to conserve space, this laminated structure is rolled into a small cylinder and placed inside a cylindrical case made of paper, metal, or plastic.

Let us first examine the effect of such a capacitor in a dc circuit. The circuit of Fig. 2-21 demonstrates the ability of a capacitor to store energy in the form of an electrostatic field. When a battery is connected to a capacitor as in Fig. 2-21, the negative terminal of the battery provides electrons to the plate of the capacitor connected to it. When an excess of electrons appears at one plate of the capacitor, the plate takes on a negative charge that repels electrons from the other plate of the capacitor. These electrons are attracted to the positive terminal of the battery. The deficiency of electrons at this plate of the capacitor leaves it with a positive charge. This difference of potential across the capacitor will equal the voltage of the battery. We say the capacitor is charged when this occurs.

When a charged capacitor is disconnected from the battery, the difference of potential across the capacitor is unchanged. Energy has been stored in the capacitor and will remain until it is allowed to discharge through an external circuit. This discharge can be seen in a large capacitor if the leads are touched together. The presence of this stored energy can also be seen if the capacitor is allowed to discharge through an ammeter, as in Fig. 2-22. The meter provides a low-resistance path between the plates. The electrons on the negatively charged plate are attracted to the positive plate through this low-resistance path. This movement of electrons constitutes a current that will be indicated by the meter. After all excess electrons on the negative plate have moved to the positive plate, the capacitor is fully discharged. There is no longer a difference of potential across the capacitor and the electrostatic field across the

Fig. 2-22. Capacitor discharging.

capacitor no longer exists. It is the ability of the capacitor to store energy in an electrostatic field that enables it to oppose changes in voltage. A capacitor, therefore, has the ability to oppose a change in voltage.

In the preceding discussion, it was stated that the capacitor was charged instantaneously. However, in the circuit of Fig. 2-23, the current is limited by resistor R. With the switch in the off position there is no current in the circuit. There is also no charge on the capacitor. The instant the switch is closed, electrons flow in the circuit. The capacitor offers no opposition to this instantaneous change in current. When an electron appears at one plate of the capacitor, it repels an electron from the opposite plate immediately. The only opposition to electron flow at this instant is the resistor. The current in the circuit is determined by Ohm's law: I = V/R. The time it takes the capacitor to charge is thereby lengthened since the current is limited. The larger the resistor, the less the current. The less current in the circuit, the longer it takes the capacitor to charge. The time it takes capacitor C to charge is also determined by the size of the capacitor. This relationship is stated by the formula C = QV. C is measured in farads; Q indicates the quantity of charge measured in coulombs; while V represents the voltage (potential difference) across the capacitor. Notice that with a given voltage (V) across the capacitor, the larger the capacitance (C) the greater the charge (Q). The charge (Q) measured in coulombs indicates the quantity of electrons on one plate of the capacitor. Therefore, the larger the capacitance, the longer it takes the capacitor to charge. The length of time it takes the capacitor to charge to 63% of the applied voltage is represented by the formula T = R × C. This value is called the time constant of the RC circuit. If R equals 1 megohm (1 million ohms) and C equals 1 microfarad (one-millionth of a farad) then RC equals 10^{-6} farads times 10^6 ohms equals 1 second. In 1 second the capacitor will charge to approximately 63% of the applied voltage. In about five time constants, or 5 seconds in this case, the capacitor in Fig. 2-23 will be fully charged.

Let us assume that in Fig. 2-23, R = 1 megohm, C = 1 microfarad, and V = 100 volts. The instant

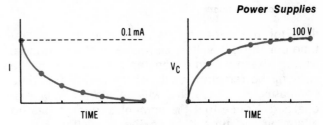

Fig. 2-24. Charging current and accumulated charge on a capacitor.

the switch is closed, the current in the circuit is 0.1 milliampere, since

$$I = \frac{V}{R} = \frac{100}{1\ 000\ 000}$$

In a short period of time, the capacitor will have assumed a charge. Let us assume that this charge has risen to 10 volts. The polarity of this voltage opposes the battery voltage. This means that the voltage drop across the resistor has decreased from 100 to 90 volts. With 90 volts across the 1-megohm resistor, the current has now decreased to 0.09 milliampere:

$$I = \frac{V}{R} = \frac{90}{1\ 000\ 000}$$

As the charge across the capacitor increases, the charging current decreases. It can be seen in Fig. 2-24 that the charge on the capacitor opposes further change.

As stated previously, in one time constant (in this case, 1 second), the capacitor has charged to 63% of the applied voltage. Since in this circuit the applied voltage is 100 volts dc, the voltage across the capacitor is 63 volts at the end of 1 second. During the next time constant, it will charge to 63% of the remaining voltage (37 volts dc). This means that at the end of 2 seconds, the capacitor will have charged to about 86 volts. In the next time constant, the capacitor charges to 63% of the remaining 14 volts. At the end of 3 seconds, the capacitor voltage is about 95 volts; at the end of 4 seconds, it is about 98.1 volts; after 5 seconds, it is 99.3 volts, or almost the full battery voltage. Notice that the capacitor actually never fully charges, although the value is so close to the applied voltage as to be indistinguishable.

A charged capacitor will also take five time constants to discharge through a resistor. If the capacitor and resistor in Fig. 2-25 are the same size as those used in Fig. 2-23, it will take the same

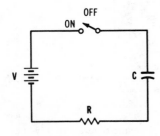

Fig. 2-23. RC circuit.

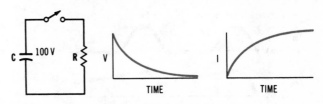

Fig. 2-25. Discharge in an RC circuit.

time for the capacitor to discharge. In the first time constant, it will discharge 63% of the voltage. During the next time constant, it will discharge 63% of the remainder, etc. At the end of five time constants, the capacitor will be almost fully discharged. Notice that the amount of charge has no effect on the time it takes the capacitor to discharge.

Fig. 2-26 shows the action of a pi-section filter in a power-supply circuit. The input to the filter circuit is the pulsating dc voltage produced by the rectifier. When the first pulsation of voltage appears across C_1, it charges to the peak voltage of this pulsation. Since there is no resistor through which C_1 must charge, it will assume this charge instantaneously. Capacitor C_1, however, must discharge through R_1 and R_L. Since C_1, R_1, and R_L are all large values, this time constant is exceedingly long. The capacitor will not discharge instantly. During the time that the rectifier output is below the peak, the capacitor (C_1) discharges. Since the time constant is long, the discharge is not great. When the next positive pulse of voltage appears, C_1 again charges. The result is that the voltage across C_1 varies very little. These variations are further filtered by the charge and discharge of C_2. Capacitor C_2 will charge to the peak voltage dropped across R_L. C_2 also discharges through R_L. This time constant is still rather long, which provides a further filtering action. Those variations that do exist in the output are called the ripple voltage. The filter should eliminate as much of this ripple as possible.

Before leaving the discussion involving the filter, a comment should be made concerning the effect of a capacitor on an ac circuit. The ratio between voltage and current (V_C/I_C) of a capacitor is called its reactance. Capacitive reactance (X_C) is measured in ohms. You will remember that a capacitor in a purely dc circuit acts as an open circuit as soon as it is fully charged; that is, the dc current becomes zero. You will notice, however, that the output of the rectifier is not a constant dc voltage; it has both an ac and a dc component. We have seen the effect of the capacitor charging to the peak voltage and maintaining this charge, due to its long time constant, during the interval between these peaks. This is also affected by its reactance. Capacitive reactance is determined by the formula:

$$X_C = \frac{1}{2\pi fC}$$

You will notice that an increase in frequency or capacitance causes a corresponding decrease in capacitive reactance. If capacitor C_1 of Fig. 2-26 is 100 microfarads, and the ac component is 60 Hz, the capacitive reactance of C_1 can be calculated by substituting these values in the formula:

$$X_C = \frac{1}{2\pi fC}$$
$$= \frac{1}{6.28 \times 60 \times 100 \times 10^{-6}}$$
$$= \frac{1}{3.768} \times 10^2$$
$$= 26.54 \text{ ohms}$$

This means that the capacitor offers a very low reactance path to the ac component of the rectifier output, which results in a very small ac voltage drop across C_1. The same results would occur at C_2, which would explain the very small ac voltage component of the power-supply output.

A COMPLETE POWER SUPPLY

The schematic diagram of a typical electronic instrument power supply is shown in Fig. 2-27. In this schematic, there are several differences from the simplified diagrams used in our discussion of power supplies. The most obvious difference is in the alternate connections provided in the primary winding of transformer T_1. Only one secondary winding is used in the actual circuit. In this case, the manufacturer probably uses this transformer in several different power supplies for similar equipment.

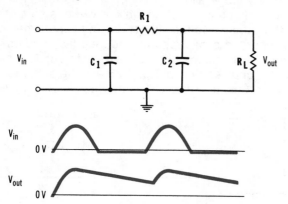

Fig. 2-26. Action of a pi-section filter.

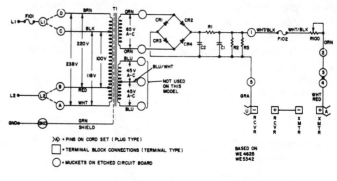

Courtesy The Foxboro Co.

Fig. 2-27. Schematic diagram of a typical electronic instrument power supply.

The power supply will develop approximately 57 volts of dc. This is determined by taking 90% of the secondary voltage and subtracting the voltage drop across two diodes. The output is also adjustable to some extent by resistor R_{100}, which is connected in series with the positive output line. The circuit is also fused in both the primary and secondary lines by F_{101} and F_{102}, respectively.

REGULATION

According to our discussion of dc power supplies thus far, it is essential that they employ a rectifier and a filter circuit. In addition to this, some power supplies have an added circuit called a regulator. The primary purpose of a regulator is to aid the rectifier and filter in providing a more constant dc voltage to the load device. Power supplies without regulation have an inherent problem of changing dc voltage values due to variations in load resistance or fluctuations in the ac line voltage. With a regulator connected in the dc output, the voltage can be maintained within a rather close tolerance of its designed output.

In its simplest form, regulation of dc can be achieved to some extent by connecting a single resistor across the power-supply output terminals (see Fig. 2-28). When variations in load resistance connected in parallel with this regulator resistor are small, some degree of regulation occurs. In this case, the regulator resistor serves as a constant load for the power supply. Under normal operating conditions, current from the power supply must pass through both the load resistor and the regulator resistor. This means that the regulator resistor serves as a fixed load regardless of the value of R_L. When the value of R_L or the input voltage changes to any real extent, this type of regulation is rather ineffective.

Zener Diode Regulators

A zener diode regulator employs a special type of semiconductor device that has an unusual conduction characteristic. When forward biased, a zener diode conducts as a conventional silicon diode. Normally, this type of diode is not used in the forward-bias direction. When reversed biased, however, it goes into conduction at a specific voltage value depending on the ratio of its doping material and silicon. Manufacturers can alter the reverse-bias conduction point, or zener voltage, when the device is being constructed. The percentage of zener voltage (V_Z) variation ranges from ±10% to ±0.1% according to the tolerance level of the selected device.

Fig. 2-29 shows a zener diode regulator connected across the dc output of a power supply. In this circuit, all of the current from the power supply must pass through the series resistor (R_s). If, for example, a 10-volt regulated output is desired, a 10-volt zener diode would be selected. The diode would be connected in parallel with R_L, as shown. As you will note, the diode is connected in a reverse-bias direction. When 10 volts appear across the diode, it goes into conduction. This voltage will be maintained for a wide range of applied voltage values.

The unique current conduction characteristic of a zener diode permits it to achieve voltage regulation. Essentially, increasing or decreasing the applied source voltage is compensated for by increasing or decreasing the current through the zener diode. This, in turn, causes the voltage drop across the series resistor (R_s) to change accordingly. An increase in line voltage would immediately cause an increase in power supply output. This would, in turn, cause a corresponding increase in current and voltage drop across R_s. As a result, the voltage across the zener diode remains at 10 volts, and the voltage across R_s changes with circuit conduction.

Similarly, a decrease in line voltage would cause a reduced current through the zener diode. This would, in turn, cause a smaller voltage drop across R_s which would maintain the voltage across the zener diode at 10 volts. In effect, this means that variations in input voltage cause a corresponding change in current which is compensated for by changes in voltage drop across R_s. Through this action, the voltage across the zener diode is maintained at a constant level.

Changes in load resistance are also reduced with a zener diode regulator. Normally, an increase in load resistance would cause a rise in dc output voltage. With a zener diode installed across the load resistor, the voltage remains at a constant value. Essentially, the zener diode compensates for this change by increasing the current through it. This, in turn, maintains the load voltage at 10 volts by the increasing voltage drop across R_s.

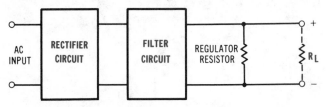

Fig. 2-28. Regulator resistor across power supply output.

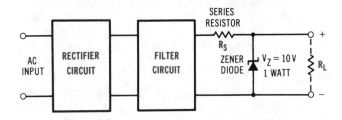

Fig. 2-29. Zener diode regulator across power supply output.

An increase in load current, caused by a smaller R_L, normally causes a decrease in load voltage without regulation. With the zener installed, however, the load voltage is maintained at the rated zener-voltage value. In this case, more current will be supplied to the load resistor, with less to the zener diode. As a result, a reduction in total current through R_s will cause less voltage drop. This, in turn, increases the load voltage to compensate for the increase in load current.

Zener diodes are normally rated by wattage and zener voltage. Wattage, for example, indicates the ability of a diode to dissipate or give off heat. In a sense, the wattage rating is an indication of the maximum current-handling capability of the diode at its rated V_z. A 1-watt, 10-volt zener diode would pass $I = W/V_z$ current without being destroyed. In this case, 1 watt divided by 10 volts equals 0.1 ampere, or 100 milliamperes, maximum current-handling capability. In practice, larger wattage ratings can be substituted in place of smaller wattage ratings. You should avoid using smaller wattage ratings when substituting any zener diode in a circuit.

VACUUM-TUBE POWER SUPPLIES

A number of vacuum-tube power supplies are still being used in electronic instruments today, although they are becoming more and more obsolete and are being superseded by solid-state units. This type of power supply is somewhat more complex than its solid-state counterpart because it necessitates a filament or heater supply in addition to the normal ac. In practice, vacuum-tube power supplies develop some rather high voltage values compared with solid-state units. This voltage is needed to supply element voltages to other vacuum-tube circuits.

As an option, if you have vacuum-tube instruments in your facility, you should study this material to become familiar with vacuum-tube circuitry. If you do not have an occasion to use vacuum-tube equipment, pass over this material and other related vacuum-tube sections.

Vacuum-Tube Rectifiers

In older power supplies, rectification is still achieved with diode vacuum tubes, which are composed of two basic elements, the cathode and the plate. The name diode means two electrodes. The cathode is made of an electron-emitting material. In this case, electron emission is produced by heat. The heat is supplied by a third element, the heater or filament. Vacuum tubes, therefore, respond to the principle of thermionic emission.

In practice, vacuum tubes are designed to operate from specific heater voltage values. The two most common values are 6.3 and 12.6 volts ac.

When the proper voltage is applied to the heater, there is sufficient current to cause the heater windings to glow and to give off a considerable amount of heat. This heat, in turn, makes the cathode temperature rise. As the temperature of a metal increases, the motion of the electrons within the metal increases. If the temperature of the cathode is high enough, some of its electrons are moving so fast that they escape from the surface of the cathode and are thus emitted ("boiled off") from the cathode. The higher the temperature, the more electrons are emitted from the surface. This temperature is limited in order to extend the life of the tube and prevent damage to it.

The emitting surface of the cathode is made of a material that is an efficient emitter, such as tungsten, thoriated tungsten, barium oxide, or strontium oxide. Electrons, thus emitted from the cathode, collect in the area around the cathode, forming a "cloud" called the space charge. Fig. 2-30 shows the symbol for a diode vacuum tube and the space charge in the region near the cathode. The polarity of the space charge is, of course, negative, since the individual electrons are negatively charged. Some of the electrons making up the space charge return to the cathode surface but other electrons emitted from the cathode immediately take their places.

When a dc voltage is applied across the diode tube so that the plate is positive and the cathode is negative, as in Fig. 2-31, electrons will flow in the circuit as indicated. This electron flow is known as plate current (I_p). The positive charge on the plate attracts the electrons that make up the space charge, causing some of them to move toward the plate. Cathode emission replaces these electrons in the space charge. The cathode-to-plate movement of electrons constitutes tube current. These electrodes are enclosed in a high vacuum inside a glass or metal case. If the tube were not evacuated, electron flow would be affected by the gas particles between the two electrodes. The higher the positive potential placed on the plate, the more

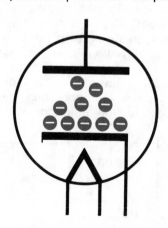

Fig. 2-30. Diode vacuum tube.

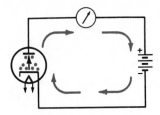

Fig. 2-31. Diode circuit with plate positive.

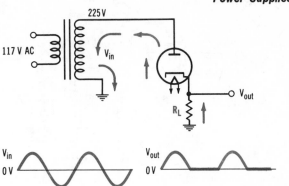

Fig. 2-33. Rectifier section of a vacuum-tube power supply.

electrons move from the cathode and are collected by the plate. These electrons also flow in the external plate circuit.

When the polarity of the voltage across the tube is reversed, as shown in Fig. 2-32, electrons will not flow. As previously described, the space charge has a negative potential. A negative charge on the plate will not attract electrons; it will actually repel these electrons and many will return to the cathode surface. Since no electron flow is present in the tube, there will be no plate current in the external plate circuit.

Fig. 2-33 is a simplified schematic of a half-wave tube rectifier circuit. One of the secondary windings of the power transformer provides an rms voltage of 225 volts to the plate of the rectifier tube. A load resistor, connected from the cathode to ground, completes the rectifier circuit.

At the beginning of an ac cycle, the plate potential rises in a positive direction. Cathode-emitted electrons are attracted to the plate, establishing tube current. The path of electron flow is as follows: (1) From ground to cathode through the load resistor. This produces a positive potential on the cathode with respect to ground. (2) From cathode to plate through the diode vacuum tube. (3) From the plate to one side of the secondary winding of the power transformer. (4) Through the secondary winding to its grounded terminal, completing the circuit. As the positive potential on the plate rises, tube current increases. This increases the current through the load resistor, which increases the voltage drop across R_L. Tube current, therefore, varies with plate voltage during the positive alternation of the ac sine wave. The IR voltage drop across R_L is then a reproduction of the positive alternation of the applied signal. The voltage across R_L is somewhat smaller than that applied to the plate during this positive half-cycle due to the resistance of the vacuum tube itself. The positive peak amplitude will be about 315

volts. When the negative alternation is applied to the plate of the diode, no electrons will flow. Hence, there will be no current in the external circuit. If there is no current through R_L, no voltage drop is developed across R_L. During the negative alternation, the output voltage (cathode voltage) remains at zero. The cycle then repeats itself.

Full-Wave Vacuum-Tube Rectifiers

Full-wave rectification is achieved with vacuum tubes in the same way that it is with solid-state diodes. Two diodes are normally housed in a single enclosure and used to achieve full-wave rectification with a center-tapped transformer (see Fig. 2-34). Note the additional heater voltage circuitry. When the power supply is first turned on, it takes approximately 30 seconds for the tube heaters to heat the cathode sufficiently to begin emitting electrons. Filament heat is an inherent problem of vacuum-tube power supplies.

Full-wave rectification can also be achieved by employing four vacuum-tube diodes or two dual-diode tubes. Fig. 2-35 shows a representative vacuum-tube bridge rectifier circuit. Note the similarity of this circuit and that of the solid-state bridge circuit. The operation is similar in nearly all respects.

A unique problem with vacuum tubes, regardless of the application, is cathode deterioration. In general, the cathode material of a vacuum tube flakes off of the cathode or in some cases becomes coated with an oxidation material. When this occurs, the active material will no longer emit a suitable quantity of electrons. As a result, the emission is reduced and the tube is described as being weak. Weak tubes must be replaced periodically in order to keep the circuit operational.

Filter Circuits

The filter circuits of a vacuum-tube power supply are nearly identical to their solid-state counterparts. In practice, the dc working voltage rating of a filter capacitor must be of a higher value to accommodate the higher operating voltages. The current in a vacuum-tube filter circuit is also

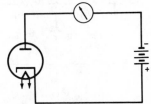

Fig. 2-32. Diode circuit with plate negative.

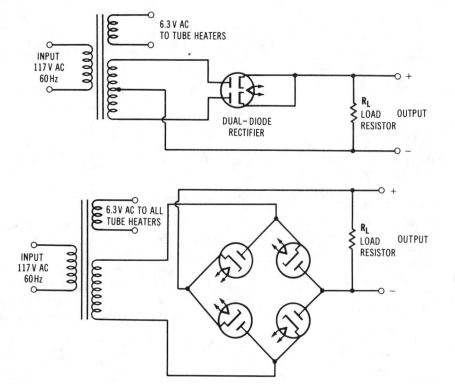

Fig. 2-34. A two-diode full-wave rectifier.

Fig. 2-35. A full-wave bridge rectifier.

somewhat less than that of a solid-state filter. Normally this means that capacitors of the vacuum-tube filter have lower capacitance values. Earlier vacuum-tube filter circuits suffered from some rather serious problems with capacitor deterioration. "Leaky" and open filter capacitors caused an abnormal rise in ripple voltage.

Complete Power Supply

Fig. 2-36 shows a typical vacuum-tube power supply such as used in older electronic instruments. The 6X4 is a seven-pin miniature dual-diode tube. Note the filter circuit and voltage divider network that provides different dc output voltages.

SUMMARY

Electronic instruments have a variety of specific functions that must be performed in order to achieve a particular operation. The power supply is responsible for the function of rectification.

A common source of input voltage to a power supply is ac. Alternating current moves first in one direction and then in the other for a short period of time. This is described as a cycle of ac. Frequency, which is the number of cycles that occur in one second, is expressed in hertz (Hz). Alternating current is produced by rotating a coil of wire in a magnetic field.

Transformers are used in power supplies to step up or step down the applied voltage. The turns ratio of a transformer determines its ability to produce a change. The output voltage and current of a transformer are inversely related.

Rectifiers are used to change ac to dc in a power supply. A two-electrode device known as a diode is used to achieve rectification. A silicon diode is made of p-type and n-type material connected together. When forward biased a diode will conduct current easily. Reverse biasing restricts current to a negligible value.

Half-wave rectification clips off an alternation of the applied sine wave, which results in a single-

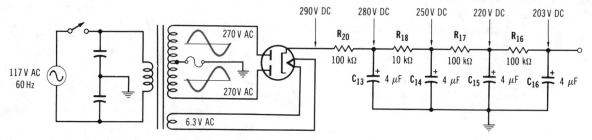

Fig. 2-36. A typical vacuum-tube power supply.

directional output current. Full-wave rectification transposes both alternations of an applied sine wave into a single-directional output. Full-wave rectification is achieved with two diodes and a center-tapped transformer, or with four diodes in a bridge arrangement.

The pulsating dc output of a rectifier must be filtered in order for it to be usable in a power supply. Filtering removes the ripple and changes the output into a constant dc voltage value.

A capacitor is two or more metal plates separated by a dielectric or insulating material. A capacitor has the ability to hold an electric charge. After it has been charged it has the ability to oppose a change in voltage. The fundamental unit of capacitance is the farad, with practical values measured in microfarads or picofarads.

When a capacitor, resistor, and a capacitor are connected together, they form a pi-section filter circuit. In operation, the first capacitor will charge, then discharge through the resistor and the other capacitor a short time later. This sequential action produces good filtering.

Regulators are frequently used in power supplies to aid the rectifier and filter in providing a more constant dc voltage to the load device. Zener diode regulators are used to produce a constant output voltage across the load resistor of a power supply. These diodes maintain a constant voltage through variations in current through a series resistor. Zener diodes are rated according to their wattage and V_z values.

Vacuum-tube diodes are used in the power supplies of older electronic instruments. These devices operate on the principle of thermionic emission. Electrons are released from a cathode that is supplied heat by a heater or filament. A positive voltage on the plate will attract electrons emitted from the cathode. Vacuum tubes may be used in half-wave, full-wave, and bridge rectifiers. As a rule, the dc output of a tube power supply is generally larger than that of a solid-state supply.

CHAPTER 3

Amplifiers

INTRODUCTION

Electronic instruments must perform a variety of basic functions in order to accomplish a particular operation. An understanding of these functions is essential to understanding the operational theory of the instrument. In this chapter, we will investigate the amplification function. Amplifiers are devices that produce a change in signal amplitude.

Amplification refers to the process of accepting a weak signal and increasing its amplitude to a higher level. In general, amplifiers employ an active device such as a transistor, vacuum tube, or integrated circuit. The amount of amplification achieved by the active device is called gain. Gain is the ratio of the output signal amplitude to the input signal amplitude. Ideally, the output signal should only reflect a change in signal amplitude. Improper component selection, voltage values, or circuit design, however, may cause the output to be somewhat distorted. The amount of distortion permitted is largely determined by the application of the amplifier.

In order for an active device to achieve gain it must have a source of operating energy and a signal to process. Direct current is used as the operating energy source for this type of device. Amplifiers in portable instruments usually receive their operating energy from chemical cells or batteries. Rectifier power supplies provide operational energy for most of the electronic instruments used in industry today.

When a signal is applied to an amplifier it causes the operating energy source to change somewhat. A small change in signal strength will, for example, cause a rather significant change in the output operating energy. The signal source may be either ac or dc and still be processed through the amplifier.

INTEGRATED-CIRCUIT AMPLIFIERS

Within the last few years integrated circuits, or ICs, have practically taken over all small-signal linear amplifier applications in electronic instruments. This type of active component contains a large number of transistors, diodes, and resistors built into a single unit. A device of this type simply requires input connections, output terminals, and an energy source. When an electrical signal is applied to the input, an amplified version of the signal appears at the output. The circuitry of this device is practically all self-contained. Fig. 3-1 shows the manufacturer's data sheet for a typical IC, with a schematic diagram of its internal components, and pin connection diagrams. This particular IC is housed in four distinct kinds of packages.

The triangle-shaped symbol is commonly used to indicate the amplification function. The point of the triangle is the output and the flat side with two leads is the input. The plus sign indicates the noninverting input and the negative sign indicates the inverting input.

The internal structure of an actual IC is quite small and very complex. It cannot be repaired when it fails to operate properly. The defective IC is simply removed from the circuit and replaced. Testing is achieved by observing the amount of gain achieved at the output of the device with respect to its input.

A large number of ICs being used today are called *operational amplifiers* or simply *op amps*. The LM741 of Fig. 3-1 is an amplifier of this type. The gain of this op amp can be controlled externally by connecting feedback resistors between the output and the input. A number of different amplifier applications can also be achieved by selecting different feedback components and combinations.

49

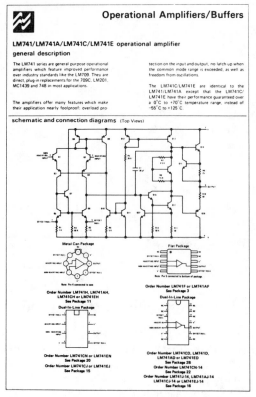

Fig. 3-1. Schematic and connection diagrams for a typical IC.

Fig. 3-2 shows an LM741 IC connected without feedback resistors. When connected in this manner the IC may achieve a gain of 100 000. This type of circuit construction represents the open-loop characteristic of an op amp. When the input voltage of this amplifier is zero the output voltage will also be zero. If an input signal is applied, the output signal will rise to a value not exceeding the source voltage. If the output rises to the source voltage the op amp is said to be saturated. An input signal of only a few millivolts is needed to cause an op amp to reach saturation.

A closed-loop circuit configuration of an op amp

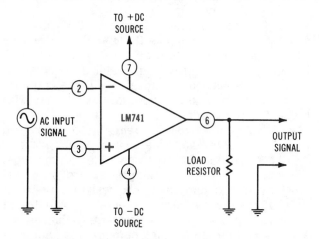

Fig. 3-2. Op amp connected for open-loop operation.

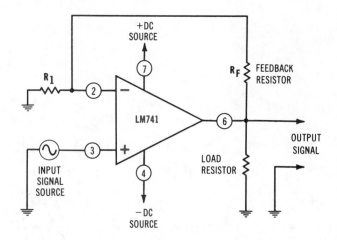

Fig. 3-3. Closed-loop IC op-amp circuit.

connected as a noninverting amplifier is shown in Fig. 3-3. In this configuration, the gain of the amplifier is determined by the resistance ratio of R_1 and R_f. Gain or voltage amplification of a noninverting IC op amp is expressed by the formula:

$$A_v = \frac{R_1 + R_f}{R_1}$$

where,

    A_v is the voltage amplification,
    R_1 is the input resistance,
    R_f is the feedback resistance.

If the input resistance of this op amp is 1000 ohms and the feedback resistor is 99 000 ohms, a voltage gain of 100 can be achieved. With a device of this type the circuit designer can simply select a combination of resistors that will permit a desired amount of amplification to be achieved.

A single IC amplifier and its associated components is commonly called a *stage* of amplification. Each amplifier stage is, therefore, designed to achieve a certain amount of gain. When two amplifier stages are connected together so that the output of the first feeds the input of the second, more gain is achieved. The total gain of a two-stage amplifier is then the product of the individual stage gains. In amplification applications that necessitate a great deal of gain, two or more IC stages may be connected together. Amplifiers of this type are often described as being *cascaded*. A number of IC designs are available today that contain four or more op amps similar to the LM741 in a single package. Extremely high gain capabilities and multifunction applications can be easily achieved with a single IC chip of this type. In the future, practically all small-signal amplifiers will be built on single IC chips.

BIPOLAR TRANSISTOR AMPLIFIERS

A single bipolar transistor amplifier is somewhat more complex to use and understand than is an

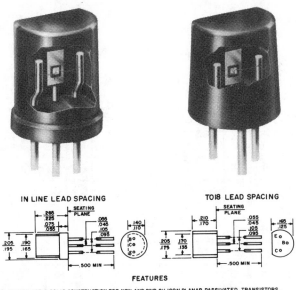

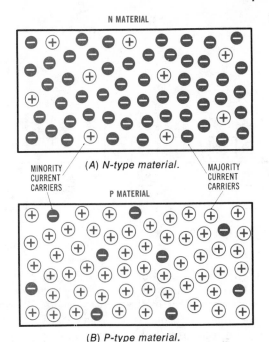

(A) N-type material.

(B) P-type material.

Fig. 3-5. The electron/hole content of n- and p-type materials.

IN LINE LEAD SPACING   TO18 LEAD SPACING

FEATURES

● MONOPLASTIC SOLID CONSTRUCTION FOR NPN AND PNP SILICON PLANAR PASSIVATED TRANSISTORS.

● PATENTED ASSEMBLY AND MOLDED ENCAPSULATION.

● CAPABILITY TO WITHSTAND EXTREME LEVELS OF MECHANICAL ACCELERATION, SHOCK AND VIBRATION.

● DEMONSTRATED MOISTURE RESISTANCE UNDER LONG TERM HUMIDITY AND TEMPERATURE.

● TEMPERATURE CYCLING CAPABILITY DEMONSTRATED OVER 300 CYCLES.

● PROVED LONG TERM OPERATING LIFE RELIABILITY.

● LEAD SOLDERABILITY.

Courtesy General Electric Co.

Fig. 3-4. The internal workings of a plastic encapsulated bipolar transistor.

IC op amp. The op amp simply uses an input, an output, and feedback loop resistance to produce amplification. Such things as forward and reverse biasing, coupling, element current, and component selection are built into the IC structure and cannot be altered. In a discrete bipolar transistor amplifier, all of these things are directly influenced by external component selection. Component selection is, therefore, very critical in the operation of this type of amplifier. The internal structure of a bipolar transistor is, however, rather simple compared with that of an IC op amp (see Fig. 3-4).

In the following discussion of bipolar transistor amplifiers, a great deal of the crystal theory of the diode is directly applicable to the transistor. In order to avoid repetition of diode theory, we will approach the subject from a slightly different point of view using the same general ideas. This will hopefully reinforce the diode theory and present some new transistor theory in a slightly different manner.

Crystal Structure

A large majority of bipolar transistors in use today are made by starting with a pure form of silicon. Impurity elements added to silicon when it is being processed cause it to be transposed into n-type or p-type materials. An n-type material has an abundance of extra electrons in its structure that are not covalently bonded with other atoms. These electrons constitute the majority current

carriers in an n-type crystal. In addition to this, there are also a few holes, or electron voids, in the covalent bonding of the structure. In an n-type material, holes constitute the minority current carriers. Fig. 3-5A illustrates the electron/hole content of a piece of n-type material.

The p-type material of a transistor is made from silicon with added element impurities such as boron (B), gallium (Ga) or indium (In). These elements have only three electrons in their valance band or outer shell. Mixing a trivalent element with silicon upsets the normal covalent bonding structure. As a result of this added impurity, a p-type of silicon material has an abundance of electron holes that do not take place in covalent

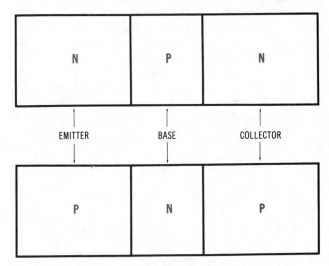

Fig. 3-6. Element names and structures of npn and pnp transistors.

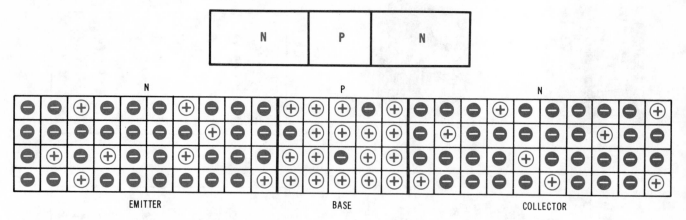

Fig. 3-7. "Egg-crate" concept of the electron/hole structure of transistors.

bonding. These holes constitute the majority current carries of a p-type crystal. Any extra or free electrons that exist in the p-type material constitute the minority current carries. Fig. 3-5B illustrates the electron/hole content of a piece of p-type material.

The general classification, *bipolar* transistor, refers to a device that employs two diode junctions formed into a sandwich-like structure. Transistors of this type are made of npn or pnp crystal formations. In this type of structure, the crystals are laminated together into a permanent unit. Fig. 3-6 shows these two structures with the element names normally associated with the respective n- and p-type materials.

Current and Biasing

To simplify our understanding of the electron/hole structure of transistors, let us think of the bipolar transistor as a three-section "egg crate." Fig. 3-7 illustrates the comparison. Consider the eggs as electrons (n-type charges) and the empty spaces as holes (p-type charges). Notice that the two n-type sections have many eggs (electrons) but very few holes. The p-type section in the middle is the opposite; it has many holes but very few electrons. If someone took several electrons from the n-type section (marked emitter) and placed them in the p-type section (marked base), what would happen? Moving the electrons from the emitter to the base would increase the number

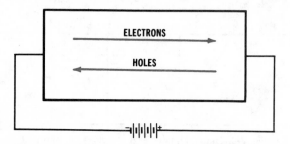

Fig. 3-8. Movement of electrons and holes in a semiconductor.

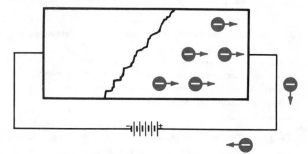

Fig. 3-9. Electrons attracted to positive battery terminal.

of holes in the emitter and reduce the number of holes in the base. So, in effect, what would happen is this:

1. Electrons would move from emitter to base.
2. Holes would move from base to emitter.

The purpose of this comparison is to make sure we realize that holes and electrons move in opposite directions in a transistor, as shown in Fig. 3-8. In the explanations that follow, the egg-crate idea will make it easier to visualize the movements of holes and electrons.

Fig. 3-9 shows that the electrons (negative charges) are attracted to the positive terminal and on into the battery. For every electron that leaves the crystal, another electron enters from the negative terminal of the battery, as shown in Fig. 3-10.

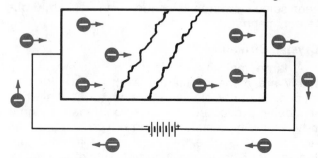

Fig. 3-10. Electrons entering crystal balance electrons leaving crystal.

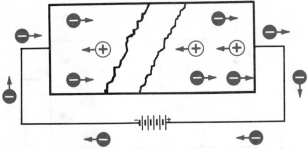

Fig. 3-11. Holes attracted to negative battery terminal.

Each electron that goes into the battery leaves a hole, as shown in Fig. 3-11. These and all other holes in the crystal are attracted to the negative terminal. The holes do not flow into the battery—they move only inside the crystal. As the holes arrive at the negative terminal of the crystal (Fig. 3-12) they are refilled by the electrons coming into the crystal from the negative terminal of the battery.

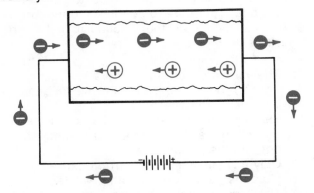

Fig. 3-12. Movement of electrons and holes in a single crystal.

Reversing the battery, as in Fig. 3-13, reverses the direction of movement of holes and electrons in the crystal. The electrons will, of course, still move toward the positive terminal, and the holes will move toward the negative terminal. Reversing the battery does not, however, have any effect on the amount of current in the circuit. This statement is valid only if we are considering one crystal element. This is a very important point in our study of transistor operation. Remember, reversing

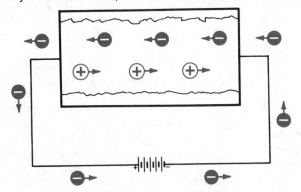

Fig. 3-13. Effect of battery reversal.

a battery connected to a single crystal element does not change the amount of current—only its direction. A current-controlling effect is realized only when two or more crystal elements are used together.

As we have learned, a bipolar transistor has three semiconductor (crystal) elements: emitter, base, and collector. All bipolar transistors have these three elements, whether they are npn or pnp type. To get an idea of how these semiconductor elements control current, let us cut the transistor in half and consider each half separately. When this is completed, we will put the halves together and apply what we have learned to the transistor as a whole.

Fig. 3-14 shows an npn transistor cut in half. Each half forms a crystal diode and is capable of controlling current. Notice that each diode section consists of an n-type and a p-type crystal. As we learned previously, an n-type crystal has a majority of negative charges (more free electrons), and a p-type crystal has a majority of positive charges (more holes). It is this difference that makes the crystal diode act as a unidirectional (one-way) current device.

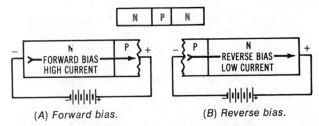

(A) Forward bias.   (B) Reverse bias.

Fig. 3-14. Forward and reverse bias.

There are obviously two ways in which a battery can be connected to a diode. The first method results in maximum current. It is called *forward bias*. The second method of connecting the battery causes the diode to act as an open circuit. This is called *reverse bias*. In Fig. 3-14A, forward bias is accomplished by connecting the negative battery terminal to the n-type crystal and the positive terminal to the p-type crystal. This produces maximum current in the diode. Reverse bias is accomplished by connecting the negative battery terminal to the p-type crystal and the positive terminal to the n-type crystal, as shown in Fig. 3-14B. This produces only negligible current in the diode.

In all cases in the following discussion, the laws of attraction and repulsion hold true:

1. Electrons inside the crystal are repelled from the negative voltage terminal and are attracted to the positive voltage terminal.
2. Holes inside the crystal are repelled from the positive voltage terminal and are attracted to the negative voltage terminal.

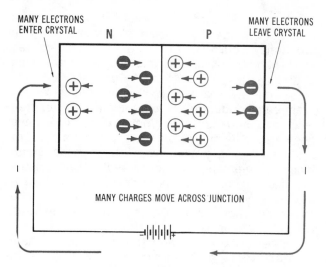

Fig. 3-15. Electron and hole movement in a forward-biased diode.

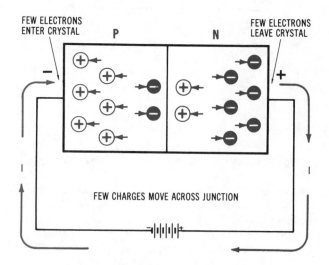

Fig. 3-16. Electron and hole movement in a reverse-biased diode.

Fig. 3-15 shows the majority current carriers in both crystals. Let us see what happens when the diode is forward biased.

1. The negative battery voltage pushes electrons in the n-type area toward the p-type area.
2. The positive battery voltage pushes holes in the p-type area toward the n-type area.
3. Because of the force applied by the voltage source, the electrons and holes penetrate the junction between the two crystals.
4. Electrons getting through to the p-type area are quickly attracted to the positive terminal and move through the conductor to the battery.
5. Every electron that moves out of the crystal leaves a hole.
6. Positive voltage pushes these holes into the n-type area.
7. The holes getting into the n-type area are attracted toward the negative terminal.
8. Electrons from the battery fill the holes arriving at the negative voltage terminal.
9. Note the fact that for every electron leaving the crystal, another one moves into the crystal.
10. It should also be noted that current consists of majority current carriers.

The majority current carriers offer a good path for current. Since there is a large number of current carriers completing the path through the diode junction, a large current occurs.

Now let us examine the other half of the transistor—the pn diode of Fig. 3-16. Notice that the negative terminal of the battery connects to the p-type crystal and the positive terminal connects to the n-type crystal. This is a reverse-bias connection. With reverse bias, the minority current

carriers in each crystal are forced to the center of the diode. Therefore, only the majority current carriers can penetrate the junction to get to the opposite side.

1. The minority current carriers, or electrons, in the p-type area penetrate the junction to get into the n-type area.
2. The minority current carriers, or holes, in the n-type area penetrate the junction to get into the p-type area.
3. Only a small number of current carriers complete the path between both crystals.
4. The few electrons that go from the p-type area into the n-type area move toward the positive terminal.
5. These few electrons go to the battery leaving a few holes behind.
6. Since only a few electrons have left the crystal, only a few electrons can enter the crystal from the negative voltage side to fill the holes.

The important thing to remember is that with reverse bias only a few current carriers exchange places between the two crystals. These are the minority current carriers. A very minute current occurs due to this action. This means that the reverse-biased junction offers a very high resistance to current.

BIPOLAR TRANSISTOR ACTION

We have discussed how current carriers flow in the individual crystal elements of a bipolar transistor. We also have considered the operation of crystal diodes biased in the forward direction and in the reverse direction. We are now going to consider what occurs when crystals are combined to form a transistor.

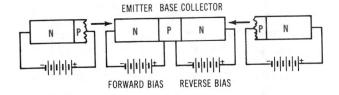

Fig. 3-17. Forward and reverse bias in a transistor.

Putting the forward- and reverse-biased diodes back into the transistor, as shown in Fig. 3-17, we notice that the emitter-to-base (np) diode section is forward biased, and the base-to-collector (pn) diode is reverse biased. Transistors are always biased in this manner. This bias method is used for both npn and pnp transistors, as shown in

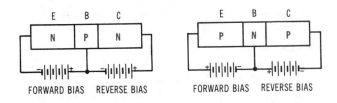

Fig. 3-18. Bias for npn and pnp transistors.

Fig. 3-18. Remember:

1. Emitter-to-base is always forward biased.
2. Base-to-collector is always reverse biased.

Let us now consider how the charges move in an npn transistor. Consider the action of the npn transistor shown in Fig. 3-19. The transistor is correctly biased with forward bias on the emitter-to-base junction and reverse bias on the base-to-collecter junction. The charge movement is as follows:

1. The bias voltage on the emitter-to-base combination causes electrons to move toward the base crystal. The base is a much thinner [about 0.001 inch (0.0254 millimeter)] crystal than either the emitter or collector crystals. Therefore, since the electrons are moving at a tremendous rate of speed, most of them (actually 95% to 99%) pass through the thin base crystal and go to the collector. The few electrons (from 1% to 5%) that do not penetrate the base are attracted to the positive voltage on the base. These few electrons result in a very small base current.

2. The great number of electrons that go to the collector are attracted to the positive terminal of the battery. These electrons leaving the collector to enter the battery make up the collector current. Every electron that moves out of the collector leaves a hole that is forced in the opposite direction by the positive collector voltage. The holes penetrate the thin base crystal and go to the emitter. Every hole that reaches the emitter is filled by another electron, which proceeds from emitter to collector.

The idea of the current-controlling action of a bipolar transistor should now become apparent. In other words, if the base is made more positive, the collector current increases. If the base is made less positive, the collector current decreases. The most important thing to remember about the npn transistor is that the forward-biased emitter-base junction controls the amount of collector current. A little later we will see how a weak signal voltage can be superimposed on the forward-bias voltage. Transistor action then produces an amplified signal in the collector circuit.

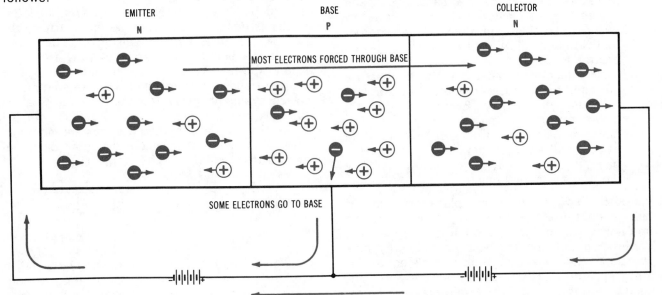

Fig. 3-19. Charge movement in an npn transistor.

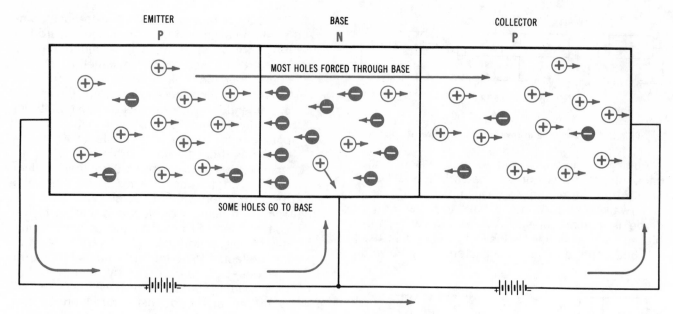

Fig. 3-20. Charge movement in a pnp transistor.

Fig. 3-20 shows the charge movement in a pnp transistor. The principles of operation for the npn and pnp transistors are basically the same; the major difference is in the crystal arrangements. The charge movement in a pnp transistor is as follows:

1. Forward bias on the emitter-base junction causes holes in the p-type material to move toward the base. Most of the holes penetrate the thin base crystal and enter the collector area.

2. As the holes arrive in the collector area, they are filled by electrons coming from the negative collector terminals. These electrons, because of the positive potential on the emitter, move through the thin base crystal and proceed to the emitter. There are also a few electrons coming from the negative base potential to fill the few holes that didn't get from emitter to collector.

3. All the electrons that go to the emitter are attracted to the positive emitter terminal and flow into the battery. Every electron that moves out of the emitter leaves a hole behind. All the holes left behind move from the emitter and most of them pass through the base crystal to the collector. A very few move to the base terminal.

Considering the foregoing explanation, it should be apparent that the major difference between npn and pnp transistor operation is in the current carriers. In the npn transistor the majority current carriers are electrons, holes, and electrons in their respective crystal material. In the pnp transistor, the majority current carriers are holes, electrons, and holes in their respective crystal

material. Again, the important point to remember is that the forward bias on the emitter-base junction controls the collector current. Thus, a small signal voltage connected in series with the emitter-base junction produces an amplified signal at the collector. We will go into this action in more detail when we analyze the operation of a basic amplifier circuit.

BIPOLAR TRANSISTOR CIRCUITS

Fig. 3-21 shows the two schematic symbols used for bipolar transistors—one is for the npn and the other is for the pnp transistor. Both symbols show the emitter, base, and collector. The collector and emitter are drawn at an angle to the base in both symbols. Notice, however, that the emitter is designated by the use of an arrow, but the collector does not have an arrow. The direction of the emitter arrow points away from the base in the npn transistor, but toward the base in the pnp transistor. Remember that the emitter arrow always points away from the direction of electron flow.

Now that we have learned how current moves in a transistor, we shall consider the three basic bipolar transistor circuit arrangements.

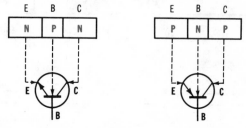

Fig. 3-21. Schematic symbols for bipolar transistors.

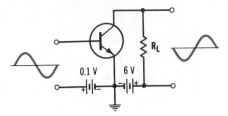

(A) Common-emitter circuit.

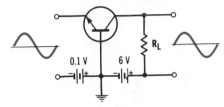

(B) Common-base circuit.

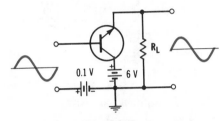

(C) Common-collector circuit.

Fig. 3-22. The three basic bipolar transistor circuit arrangements.

1. The common-emitter circuit of Fig. 3-22A has a common connection between the base input and the collector output.
2. The common-base circuit of Fig. 3-22B has the base commonly connected to the emitter input and the collector output.
3. The common-collector circuit of Fig. 3-22C has the collector commonly connected to the base input and the emitter output.

To illustrate how each of the three types of transistor circuits works, we will discuss each one separately. The circuits of Fig. 3-22 will be used. The input to each circuit will be achieved by a transformer.

Fig. 3-23 might cause some confusion if you recall that the base-to-collector junction must be reverse biased. Actually, if the schematic is studied, it will be noticed that the base-to-collector

junction really is reverse biased at a value of −5.9 volts. This can be understood by visualizing the +6 volts of the collector battery as appearing at the collector, while the +0.1 volt of the bias battery appears at the base. The difference between the two voltages is 5.9 volts. The base is negative with respect to the collector, as required for the reverse-bias condition.

In the steady-state condition, the positive 0.1 volt on the base will cause a certain amount of current to move from the emitter to the collector. The actual amount of this current will depend on the characteristics of the transistor involved. We know from our previous discussion of transistor action that most of the current passes to the collector and very little leaves through the base. The current in the collector circuit passes through the load resistor on its way back to the battery. This current through R_L produces a voltage drop across the resistor. The output for this circuit in Fig. 3-23 is taken between the collector and the emitter.

Consider now how a signal applied to the base affects the circuit. As the input signal rises in the positive direction, the signal voltage causes the transistor to conduct more. When the input signal drops in the negative direction, the signal subtracts from the +0.1-volt bias voltage. This causes the base to become less positive with respect to the emitter; therefore, fewer electrons (less current) move from emitter to collector. This changing collector current will develop a signal across R_L. This amplified output is coupled through C_C to the output.

We recall that in the common-emitter circuit, it was difficult to detect the presence of reverse bias on the base-to-collector junction. In Fig. 3-24, the common-base circuit, the reverse-bias condition is quite obvious. In this circuit, during the steady-state condition shown in the diagram, the +0.1-volt forward bias on the base-emitter junction would produce a small amount of current. Here, as before, the magnitude of this current would be dependent on the characteristics of the transistor.

When an ac signal voltage is applied to the circuit, the positive alternation subtracts from the 0.1-volt battery potential. The emitter, therefore, becomes less negative, or more positive. This reduces the forward-bias voltage, and the transistor current decreases, resulting in an increase in the output voltage. The emitter becomes more nega-

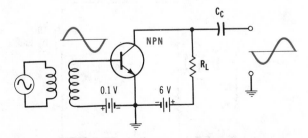

Fig. 3-23. Signals in common-emitter transistor circuit.

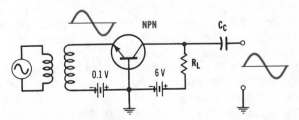

Fig. 3-24. Signals in common-base transistor circuit.

tive during the negative alternation of the input signal. This increases the collector current. A corresponding decrease in the output signal results.

In the common-collector circuit, we have a bias condition similar to that used in the common-emitter circuit. Notice in Fig. 3-25 that the reverse bias on the base-to-collector junction is accomplished by the +0.1-volt battery making the base 5.9 volts less positive (negative) with respect to the collector. Thus, the reverse-bias condition is accomplished. In this circuit, the input is applied to the base and the output is taken from the emitter. The collector is the common element. In other words, it is associated with both the input and output circuits.

During the positive portion of the input signal, the forward bias between the emitter and base increases, producing more current in the transistor. This current moves through the load resistor (in the emitter circuit), causing an output to be developed. When the input signal swings into the negative region, it subtracts from the forward bias on the base-to-emitter junction. This decreases the amount of current in the transistor; consequently, less voltage is developed across the load resistor.

In the common-collector configuration, the output signal is less than the input signal. The output also follows the input in all respects. In this regard, a common-collector circuit is frequently called an emitter-follower amplifier. Circuits of this type are normally used to achieve impedance matching.

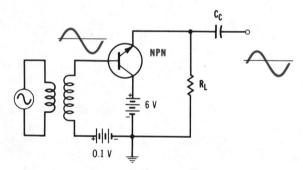

Fig. 3-25. Signals in common-collector (emitter follower) transistor circuit.

FIELD-EFFECT (UNIPOLAR) TRANSISTORS

Field-effect transistors (FETs) represent the second major type of discrete solid-state amplifying devices. These devices, unlike their bipolar counterparts, conduct current in a single piece of semiconductor material known as a channel; thus, they are also known as *unipolar* transistors. Junction FETs and metal-oxide semiconductor FETs are the two general classifications of FETs. Element names and theory of operation are significantly different than the bipolar type of transistor.

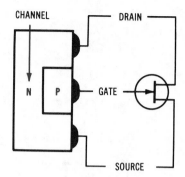

(A) N-channel JFET.

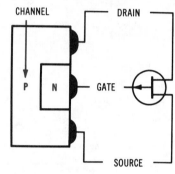

(B) P-channel JFET.

Fig. 3-26. Junction field-effect transistors.

The junction field-effect transistor (JFET) is a modification of the bipolar transistor. Structurally, JFETs conduct current through a single piece of semiconductor material called a *channel*. An additional piece of semiconductor material attached to the side of the channel is called a *gate*. Current conduction through the channel is controlled by reverse biasing of the gate. Fig. 3-26 shows representative crystal types, element names, and schematic symbols of the junction field-effect transistor.

A simple JFET amplifier circuit is shown in Fig. 3-27. In this circuit, the positive side of the battery, V_{DD}, is connected to the drain through R_D, and the negative side is connected through R_S. With these two connections made, current easily passes through the channel. A voltage drop across R_S makes the source electrode slightly positive with respect to the negative battery terminal.

When resistor R_G is attached to the negative side of V_{DD}, it reverse biases the gate-source junction.

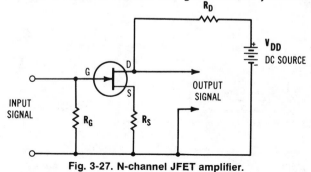

Fig. 3-27. N-channel JFET amplifier.

Action of this type is used to control conduction of the channel current. A positive signal applied to the input would reduce this bias voltage and cause increased channel current conduction. A negative signal would add to the reverse biasing of the gate and cause less channel conduction. Ac signals applied to the gate will cause a constant variation of bias voltage above and below a certain operating point. The output would then reflect this as an I_D that changes in step with the ac input signal.

When the gate bias voltage of a FET becomes high enough it will completely stop the drain current, I_D. The term *pinch-off voltage* is commonly used to describe this operating condition. This condition corresponds to the reverse biasing action of the base-emitter junction of a bipolar transistor.

Since the gate of a FET responds very quickly to changes in voltage, it is classified as a voltage-sensitive device. This represents a unique feature of the FET compared with bipolar devices, which are considered current sensitive. Because of this feature the n-channel FET has operating characteristics that are very similar to those of a vacuum tube.

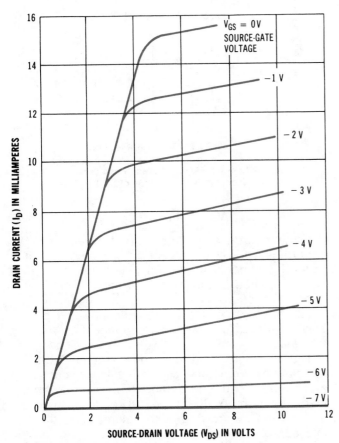

Fig. 3-28. A family of I_D/V_{DS} characteristic curves for an n-channel JFET.

JFET Characteristic Curves

A family of JFET characteristic curves is shown in Fig. 3-28. The x axis of this graph shows the voltage applied between drain and source as V_{DS}. The y axis of the graph is used to display the drain current, I_D, in milliamperes. The pinch-off voltage of this device is −7 volts dc. Operation of this particular JFET at gate control voltages between $V_{GS} = 0$ volts and $V_{GS} = -5$ volts is permissible.

The gain of a FET is commonly described by the term *transconductance,* which has the letter symbol g_m. This term refers to the input/output relationship of the device. A change in V_{GS}, for example, causes a corresponding change in output current, I_D. The formula $g_m = I_D/V_{GS}$ permits easy calculation of transconductance.

Assume now that the transconductance of a JFET amplifier having the characteristic curves of Fig. 3-28 is to be determined. If the V_{DS} appearing across the device is held at 8 volts, this will serve as a point of reference. A change of V_{GS} from −1 to −2 volts along the reference line would cause a corresponding change in I_D of 13 to 10.5 milliamperes. Entering these values into the transconductance formula will show that:

$$g_m = \frac{I_D}{V_{GS}}$$

$$= \frac{13 \text{ to } 10.5 \text{ mA}}{-1 \text{ to } -2 \text{ V}}$$

$$= \frac{0.0025}{1}$$

$$= 2500 \text{ microsiemens}$$

The transconductance of a JFET is a measure of the ease with which current carriers pass through the channel. Conductance, in this case, is the reciprocal of resistance, or $G = 1/R$. The unit of conductance is the siemens. The ease with which current carriers pass through a FET is, therefore, measured in siemens or, more commonly, microsiemens. This same term is used to show the gain of a vacuum tube.

MOS Field-Effect Transistors

Metal-oxide-semiconductor FETs (MOSFETs) are the newest variation of the unipolar transistor family. MOSFETs are designed so that the gate electrode is insulated from the channel by a layer of metal oxide. Reverse biasing of the gate-source electrodes causes a depletion of drain current carriers in the channel, like that of the JFET. The insulated-gate feature of this device can also permit an increase in drain current to take place when appropriate bias voltages are applied. This type of operation is described as the *enhancement mode* of operation. This condition of operation occurs when the gate is forward biased. JFETs, by

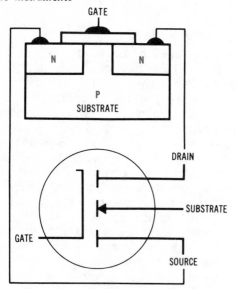

(A) N-channel enhancement MOSFET.

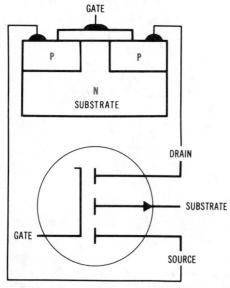

(B) P-channel enhancement MOSFET.

Fig. 3-29. Metal-oxide semiconductor field-effect transistors.

comparison, cannot be operated when the gate is forward biased because it would cause a large gate current. With the insulated-gate characteristic of the MOSFET, the polarity of the gate bias

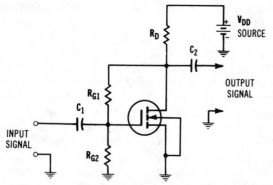

Fig. 3-30. N-channel enhancement-mode MOSFET amplifier.

voltage is not a problem. Fig. 3-29 shows representative crystal types, element names, and schematic symbols for the MOSFET.

Fig. 3-30 shows the circuit diagram of an n-channel MOSFET amplifier. This circuit is very similar to that of the JFET amplifier of Fig. 3-27. Resistors R_{G1} and R_{G2} form a voltage divider across the power source to bias the gate. The ratio of these two resistors can be altered to provide different values of V_G voltage according to the demands of the circuit application. This resistance ratio can be altered to make the gate positive for forward biasing in the enhancement mode or negative for reverse biasing in the depletion region.

Fig. 3-31 shows a drain family of characteristic curves for an enhancement-depletion type of MOSFET. Note that this device can be used in either a positive or negative gate-biasing application. Tranconductance calculations and circuit operations of the MOSFET and JFET are practically identical in all respects.

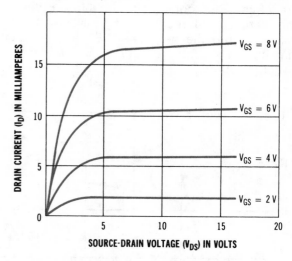

Fig. 3-31. A family of I_D/V_{DS} characteristic curves for an enhancement-depletion type of MOSFET.

VACUUM TUBES

Vacuum tube amplifiers still represent a very important type of amplifying device used in electronic instruments today. As a general rule, this type of device is found in instruments that have been in operation for some time. These instruments have proven to be reliable and will probably continue to be used for some time in the future.

The triode, or three-electrode vacuum tube, is the simplest amplitude control device. It contains a cathode, a grid, and an anode, or plate. A schematic diagram of the tube, its element names, and a cutaway drawing of the elements is shown in Fig. 3-32. Notice that the filament and cathode serve as a single unit and that the grid is wound in a helix around the cathode. The cathode is de-

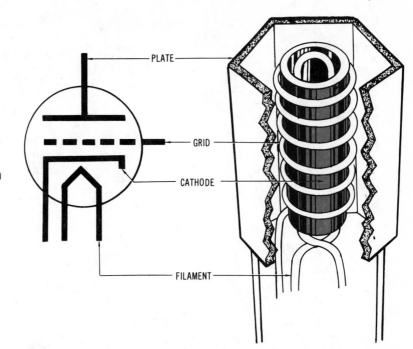

Fig. 3-32. Schematic symbol and cutaway drawing of a triode vacuum tube.

PLATE

GRID

CATHODE

FILAMENT

signed to emit or give off electrons when it is heated by the filament.

The basic principle of vacuum tube operation is called *thermionic emission*. This principle states that a number of electrons are thrown out of their orbital path around the nucleus when certain materials are heated. The cathode in this case serves as the emitting material of a vacuum tube. The filament is needed to provide heat for the cathode so that it will produce an emission. A separate filament source is needed to operate the vacuum tube.

After electrons are emitted from the surface of the cathode they must pass through the grid. This electrode serves as the control element of the vacuum tube. With no signal applied to it, the control grid normally has a voltage of negative polarity on it. Electrons are, therefore, repelled by the grid according to the amount of voltage applied. A signal applied to the grid will add or reduce grid voltage accordingly. This in turn will increase or reduce electron flow through the grid.

The plate, or anode, of a vacuum tube is designed to collect electrons that are emitted from the cathode. This electrode must be attached to the positive side of the source that feeds the cathode. As a result of these connections, electrons travel through space from cathode to anode while being controlled by the grid. Fig. 3-33 shows a typical triode amplifier circuit.

A vacuum tube is classified as a voltage-sensitive device. Changes in grid voltage, for example, are used to control the plate current, I_P, through the tube. Initially this current starts at the negative side of the source, V_{BB}, and travels through resistor R_K, the tube, resistor R_L, and back to the posi-

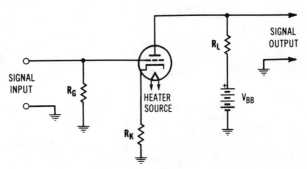

SIGNAL INPUT

R_G

R_K

HEATER SOURCE

R_L

SIGNAL OUTPUT

V_{BB}

Fig. 3-33. Vacuum-tube triode amplifier circuit.

tive side of the source. A change of I_P at any point will cause the same change throughout the circuit.

A typical family of plate characteristic curves is shown in Fig. 3-34. A small change in grid voltage will cause a rather significant change in plate current I_P. This relationship shows the transconductance, or g_m, of the tube. Expressed as a formula:

$$g_m = \frac{I_P}{V_G} \text{ with } V_P \text{ held constant}$$

where,

V_G is the change in grid voltage that causes the change in plate current I_P.

The voltage gain of a vacuum tube amplifier is also a very important characteristic. This item is called the *amplification factor*, or mu (μ), of a vacuum tube. Expressed as a formula:

$$\mu = \frac{V_P}{V_G}$$

where,

I_G and I_P are held constant,
V_G and V_P are small changes.

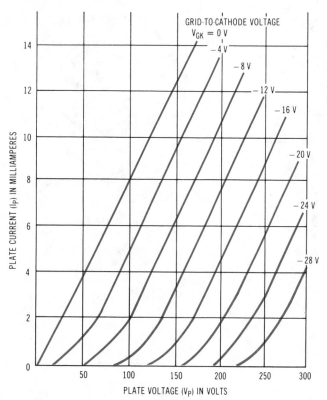

Fig. 3-34. A family of I_P/V_P characteristic curves for a triode vacuum tube.

SMALL-SIGNAL VERSUS LARGE-SIGNAL AMPLIFIERS

Practically all of the active devices that will achieve amplitude control can be classified as either small-signal or large-signal amplifiers. This particular classification indicates the location of the amplifier in the system. Initially, the input signal of a system is quite small and requires a great

Courtesy General Electric Co.

Fig. 3-35. The internal structure of a power transistor.

deal of amplification. Small-signal amplifiers are, therefore, purposely designed to accept a weak signal and amplify it accordingly. After the signal has been amplified rather significantly, it can then be used to control larger amounts of current that are fed into the load device. Large-signal amplifiers are, therefore, designed to achieve this function. These devices are substantially larger in physical size and often include special heat-dissipating connections. Fig. 3-35 shows the internal structure of a power transistor. This device is capable of controlling several amperes of collector current as compared with a few milliamperes by the small-signal transistor.

SUMMARY

Amplification is a process in which a small signal is applied to the input of an active device such as a transistor, integrated circuit, or vacuum tube, and its output amplitude is increased to a higher level.

Integrated circuits are commonly used to achieve amplification in electronic instruments today. ICs such as the operational amplifier are built on a single IC chip. Without a feedback resistor, gains of 100 000 are common. With a feedback resistor, gain is determined by a resistance ratio of the input and feedback resistors.

Discrete transistor amplifiers must take into account such things as forward and reverse biasing, component selection, and element current. Bipolar transistors have two junctions in an npn or pnp structure. Current is primarily based on the flow of majority current carriers. A conventional connection method has the emitter-base forward biased and the base-collector reverse biased. Transistors can be connected into a circuit that can have any one of the three elements common to the input and output circuitry. Common-emitter, common-base, or common-collector amplifier configurations are in use today.

JFETs and MOSFETs are unipolar transistors. These devices respond to current carriers passing through a crystal channel that has the source and drain electrodes attached to it. Control is achieved by biasing the gate element.

Vacuum tubes represent the third type of active device. Tubes respond to the principle of thermionic emission. Electrons emitted from the cathode are controlled by the voltage applied to the grid electrode. Current is from cathode to plate while passing through the grid. Vacuum tubes are generally considered obsolete today in most electronic instruments.

CHAPTER 4

Electronic Recording Instruments

INTRODUCTION

The first instrument that will be discussed in this presentation is the electronic recorder. This type of instrument, like others, must employ a a number of basic electronic functions in order to operate. Power supplies, voltage amplifiers, dc to ac converters, and servomotor control circuits are some of the representative electronic functions that make this instrument operate.

In general, electronic recording instruments are designed to provide a type of graphic display of variations in a particular quantity being measured. In some applications this may be a graphic instrument that places information on a circular chart or a strip chart that responds to control of a servomotor unit. In addition, recorders may also display the measured quantity on a hand-deflection instrument. Fig. 4-1 shows a single-point chart recording instrument that records on a 6-1/2-in (165-mm) calibrated strip chart. A similar instrument designed for recording on a 3-in (76-mm) round chart is shown in Fig. 4-2. A hand-deflecting indicator is shown in Fig. 4-3 for comparison. This type of indicator displays the measured variable on a 23-in (584-mm) scale.

Electronic recording instruments usually have the capability of performing some type of control function in addition to their information gathering function. Instruments of this type are normally described as controllers or recording controllers. In this regard, a controller is ultimately used to alter a process variable or control a system function. This chapter is primarily concerned with the re-

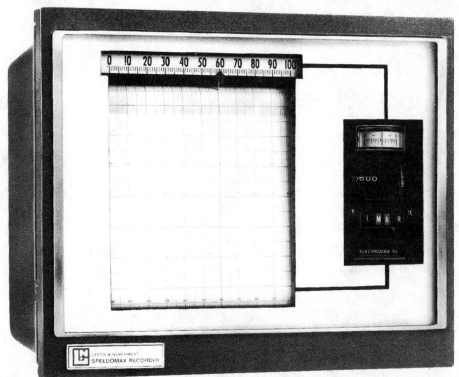

Fig. 4-1. A single-point strip chart recording instrument.

Fig. 4-2. A circular chart recording instrument.

Courtesy Leeds & Northrup Co.

cording-indicator function, while some of the later chapters will be concerned with controller operation.

Electronic recording instruments are basically either of the null type or of the galvanometer type. A null instrument generally responds to comparisons between the measured input and a set-point input applied to the instrument. Balance may be achieved manually or it may be self-balancing. Self-balancing instruments seem to be universal throughout the industry today.

FUNCTIONAL UNITS

Fig. 4-4 shows a block diagram of the functional units of an electronic recorder of the servomotor type. The major blocks of this diagram that are of concern in this chapter are the power supply, converter or chopper, and voltage amplifiers. These functions will be discussed in rather general terms that could apply to the basic circuitry of nearly any recording instrument.

Power Supply

The power supply of an electronic recording instrument is primarily responsible for providing the voltages needed to make the unit operate. A very important part of the power supply is used to develop the dc voltage that is supplied to all of the active components of the recorder. Rectifiers, filters, and voltage regulators are utilized to achieve this function. In addition to this, some instruments employ a constant-voltage unit

Courtesy Leeds & Northrup Co.

Fig. 4-3. A hand-deflection recording instrument.

that provides stable dc voltage values for reference standards. This unit commonly replaces the standard cell. The ac voltage developed by the power supply is applied to the chart motor of in-

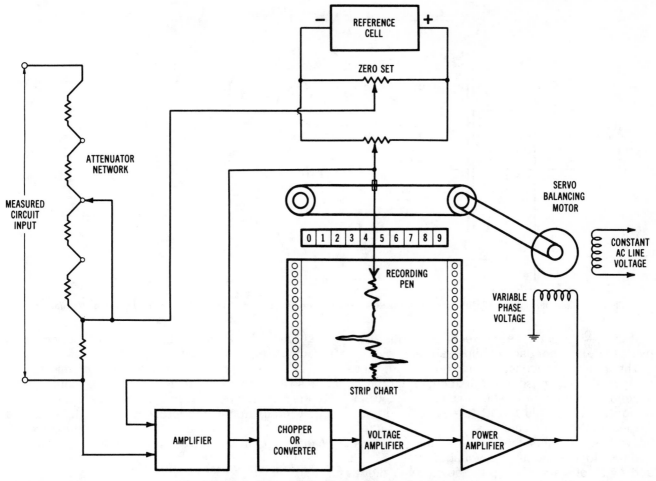

Fig. 4-4. Block diagram of an electronic recording instrument.

struments that employ one. Low voltage ac is also supplied by a transformer to the chopper or converter components. This voltage is generally developed by a separate winding on the power supply transformer.

Converter Circuits

The primary purpose of the converter circuit is to change any dc input developed by the measuring circuit into an ac voltage. This conversion is necessary in order to use a stable, high-gain voltage amplifier. These circuits also help form the ac signal that is to be the input to the voltage amplifier.

Amplifier Unit

The amplifier unit of a typical recording instrument achieves combined functions of voltage and power amplification. In many cases, the entire amplifier assembly is described as a servomotor control amplifier unit. It is generally built on a single printed-circuit board or card for easy installation and replacement. This unit is specifically responsible for receiving an error signal, which is the difference between the measuring input and the feedback circuit, and amplifying it to drive the motor unit.

A major part of the amplifier unit involves several stages of voltage amplification. This part of the instrument is primarily responsible for amplifying the input signal to a level that is large enough to drive the power amplifier.

The power amplifier of the recorder could more accurately be called a motor control circuit. It produces a signal with a current high enough to drive the servomotor. In a sense the power amplifier is a rather low-resistance transistor that permits a larger current than can be achieved by a small-signal amplifier. A power amplifier requires a high amplitude input signal in order to control the current to the motor.

POWER SUPPLY CIRCUITS

Fig. 4-5 shows a simplified schematic diagram of a representative power supply used in an electronic recording instrument. In this particular circuit, 120 volts rms, 60 Hz serves as the power supply input. Through normal transformer action this voltage is stepped down to a desired value

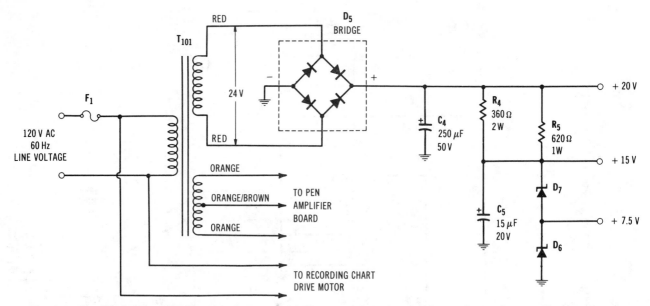

Fig. 4-5. Simplified schematic diagram of a representative power supply used in an electronic recording instrument.

that is used to energize the rectifier. In this case, two separate secondary windings are employed.

The red-wire secondary winding of the transformer is used to supply the rectifier circuit. A secondary voltage of 24 volts rms is applied to the bridge rectifier. This develops +20 volts dc with respect to the common chassis ground. This specific voltage supplies the pen actuating motor. In addition to this, regulated voltages of +15 volts and +7.5 volts are developed across zener diodes D_6 and D_7. Resistor R_5 serves as the series dropping resistor for these voltages. In this case, the zener diodes are connected in series and have the same V_z rating. These two low voltages are used to supply the transistors of the circuit. Capacitor C_4 serves as an input filter for the +20-volt supply, and C_5 filters the +15-volt supply.

The orange-wire secondary windings of the transformer serve as a supply for the pen amplifier of the recorder. In this particular unit, the amplifier has a split power supply of +15 volts and −15 volts for the IC voltage source built on the amplifier circuit board. Through this alternate power source, common power-supply noise problems are minimized. The split dc power supply will be dis-

cussed in a later chapter in which IC operational amplifiers are utilized.

The recording chart drive motor of this unit is also remotely located but derives its source voltage from the power supply. In this particular circuit the chart motor is energized by the 120-volt line voltage. The motor and transformer primary winding are both fused by F_1 for circuit protection.

A Constant-Voltage Source

Fig. 4-6 shows a constant-voltage source that is used in Honeywell Electronik 15 recorders for the potentiometer standard of a bridge circuit. Transformer T_1 has 120-volts, 60-Hz line voltage applied to it, and a dc voltage of 1.029 volts, 4.2 volts, or 5.064 volts is developed at the output according to the values of selected resistors R_m and R_c. The supply is actually a half-wave rectifier with C_1 forming a capacitive input filter. Regulated output voltage is achieved by zener diodes D_1 and D_2. This specific power supply is built on a compact circuit board and is housed in a metal container for shielding purposes. The current output of this supply ranges from 6 to 8 milliamperes at its rated output voltage.

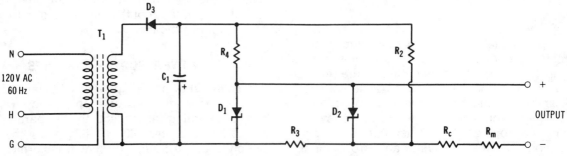

Fig. 4-6. A constant-voltage source used with electronic recording instruments.

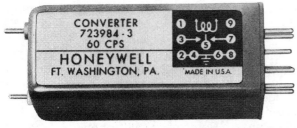

Courtesy Honeywell, Inc.

Fig. 4-7. A mechanical dc to ac converter.

DC TO AC CONVERTERS

Today, dc to ac converters are a common part of nearly all electronic recording instruments. The responsibility of this circuit is simply one of changing the measured input signal, which is dc, into a usable form of ac. A mechanical converter, such as the one shown in Fig. 4-7, is often used to achieve this function.

Components

Only three functional components are utilized in the converter circuit of a recorder. This includes the converter itself, which is sometimes called a chopper, the input transformer, and the filter. The chopper and input transformer work together to change dc to ac. The filter is used to suppress unwanted noise and voltages that may be superimposed upon the dc input signal.

Operation of the Chopper

The chopper consists of a drive coil, a vibrating reed, and a set of contacts. The chopper is shown schematically in Fig. 4-8.

Terminals 4 and 5 connect the drive coil of the chopper to a 6.3-volt ac, 60-Hz secondary winding of the power transformer. This ac voltage produces a changing magnetic field in the core of the drive coil. Let us assume that during the positive alternation of voltage across the drive coil terminal 4 is negative and terminal 5 is positive. Current will then conduct through the drive coil from terminal 4 to terminal 5. This direction of current conduction produces a north magnetic pole at the left end of the core of the drive coil. The right end of the core becomes a south magnetic pole. During the negative alternation of an ac voltage applied to the drive coil the direction of current is reversed. This reversal of current produces a south magnetic pole on the left end of the drive coil. Since the frequency of the voltage connected to the drive coil is 60 Hz, the polarity of the magnetic field also reverses 60 times per second.

The vibrating reed is actually the center contact of a single-pole, double-throw switch. Attached to the end of this thin metal strip is a permanent magnet. This permanent magnet is placed near one end of the drive coil. In Fig. 4-8, if this end of the drive coil is the north pole of the electromagnet produced by the coil, the vibrating reed will move upward. This movement is the result of the attraction between the south pole of the permanent magnet and the north pole of the drive coil, and the repulsion of the two north poles. This movement will also close the circuit between terminals 1 and 2. When the current to the drive coil reverses direction, a south pole is produced at this end of the coil. This causes a downward motion of the vibrating reed. The action is again produced due to the reaction between the pole of the drive coil and the poles of the permanent magnet. When the reed moves downward, contact is made between terminals 2 and 3. Since the drive coil voltage has a frequency of 60 Hz, the vibrating reed contacts both terminals, 1 and 3, 60 times per second.

DC to AC Conversion

The conversion of the dc voltage produced by the measuring circuits into an ac voltage incorporates the action of the chopper and the primary of the input transformer. The connections of these two components are shown in Fig. 4-9.

This schematic indicates the situation when the output of the measuring circuits is positive. Under these conditions the voltage at terminal 2 (the vibrating reed) is positive and the center tap of the input transformer is negative. As the vibrating reed moves, terminals 1 and 2 are connected. The circuit is closed through the upper half of the primary winding of the input transformer. The dc

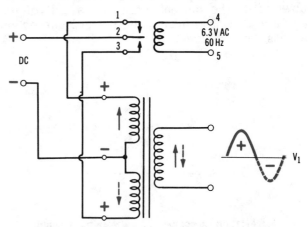

Fig. 4-9. Conversion of dc to ac with positive input.

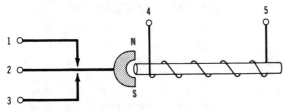

Fig. 4-8. Diagram of a chopper.

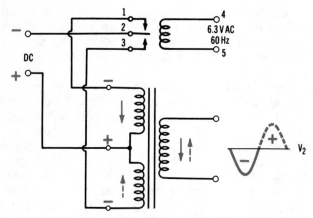

Fig. 4-10. Conversion of dc to ac with negative input.

voltage across this part of the transformer produces a current through this part of the primary. A voltage is induced into the secondary of the transformer with the polarity (positive) indicated due to this primary current. The vibrating reed then moves downward. Terminals 2 and 3 are now connected. This supplies a voltage across the lower half of the primary. The current is now conducting in the primary in the opposite direction to that described before. This current induces a negative voltage across the transformer secondary. The action of the vibrating reed produces a signal which closely approximates a square wave. The inductive action of the transformer, however, shapes this waveform so that the transformer output now closely approximates the sine wave.

Fig. 4-10 shows the polarity of the voltages when an unbalance occurs in the measuring circuit to produce a negative input to the amplifier. The vibrating reed moves upward, closing contacts 1 and 2. A voltage is then applied across the upper half of the transformer primary. Current is produced in this half of the primary that induces a negative voltage into the secondary winding. A voltage is provided across the lower half of the transformer primary when the vibrating reed moves downward. The current through this part of the primary winding produces the positive alternation across the secondary.

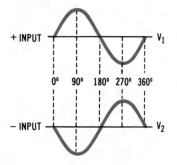

Fig. 4-11. Phase comparison between positive and negative inputs.

The difference between the ac voltage output under the two conditions should be noted. Fig. 4-11 shows this relationship. You will notice that during the positive alternation produced by a positive dc input a negative alternation is produced by a negative dc input. Anytime the positive maximum voltage occurs at the same time that the negative maximum occurs, the voltages are said to be 180° out of phase. Remember that a conductor moving in a circular path (360° rotation) between a pair of poles of a magnet produces one cycle of an alternating voltage. The time reference of an ac voltage, therefore, can be measured in degrees. The time difference between two corresponding points on two ac voltages measured in degrees is the phase difference or phase shift. You will notice that the time difference between the positive peak of V_1 and V_2 is one-half cycle or 180°. This is a 180° phase shift.

When the measuring circuits are balanced in the null position, the zero-voltage input will produce no ac signal as an amplifier input. The filter circuit is of particular importance under the zero input conditions. This filter is made up of a parallel combination of a resistor and capacitor. This circuit is shown in the complete schematic of the converter circuits in Fig. 4-12. The capacitor offers a low-impedance path to very high frequencies. Any noise which might be picked up and introduced as a signal into the amplifier should be bypassed through this low impedance path to ground.

CONVERTER CIRCUITS

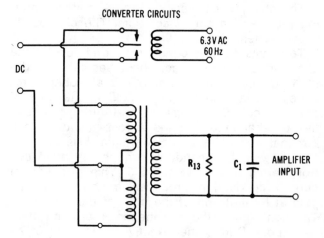

Fig. 4-12. Converter and filter.

VOLTAGE AMPLIFIERS

The block diagram of the recorder shown in Fig. 4-4 contains several distinct voltage amplifiers that are not shown specifically. These amplifiers process the ac signal from the converter through several stages of RC-coupled amplifiers. Terms such as preamplifiers, signal amplifiers, and driver amplifiers are often used to describe this part of the recorder. Amplifiers of this type are primarily responsible for most small-signal amplification applications. This type of amplification increases the amplitude of the applied signal voltage to a rather high level at the output. The most important component of a voltage amplifier is its active device, which is either a bipolar or a field-effect transistor.

Instead of discussing a particular amplifier produced by a specific manufacturer, we will approach the subject of amplification in rather general terms. The principle of operation can then apply to any one of several recorders available today. Through this approach you should be able to apply the amplification function to nearly any specific circuit application.

NPN Transistor Circuit

Fig. 4-13 shows the necessary connections of a simple npn silicon transistor connected into a common-emitter amplifier circuit. The operation of this circuit is based on the conventional biasing of a transistor, which was discussed previously. The emitter-base junction is forward biased, with the collector-base junction reverse biased. As noted, forward biasing of the emitter-base junction is accomplished by connecting the negative side of the battery (V_{CC}) to the emitter and the positive side, through resistor R_B, to the base. Collector current through load resistor R_L is controlled by variations in the forward bias voltage of the emitter-base junction.

A transistor amplifier of this type is designed to control the amount of current that passes through it. Current from the battery enters the emitter, passes through the base, and exits through the collector. Variations in output, or collector current (I_C), can be made by changing the base current (I_B). A small change in I_B causes a rather substantial change in I_C. This relationship is commonly described as transistor current gain or *beta*. Expressed mathematically,

$$beta = \frac{\Delta I_C}{\Delta I_B}$$

All of the current entering a transistor at the emitter is referred to as emitter current, or I_E.

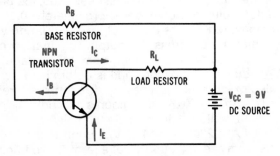

Fig. 4-13. Basic npn transistor circuit.

The output current, I_C, is always somewhat less than I_E because of the base current. Mathematically this is expressed by the formula,

$$I_C = I_E - I_B$$

PNP Transistor Circuit

The transistor amplifier circuit of Fig. 4-14 is a pnp counterpart of the previous npn circuit. The battery of this circuit is connected in a reverse direction in order to achieve proper biasing. Performance is basically the same as that of the npn circuit. Currents I_C, I_B, and I_E are represented in this diagram by arrows. The emitter current of this circuit still provides the largest current value. The composite of I_C plus I_B also equals I_E in this circuit.

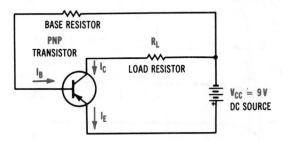

Fig. 4-14. Basic pnp transistor circuit.

Static Transistor Circuit Conditions

When a transistor is developed, the manufacturer produces operating specifications and characteristics that are used to predict its performance. A typical family of collector curves is shown in Fig. 4-15. The collector-emitter voltage (V_{CE}) is plotted on the x axis of this graph, while the collector current (I_C) is plotted on the y axis. The individual curves of the graph represent different base-current (I_B) values. The zero base-current line is omitted because it represents the cutoff condition of the transistor.

A family of collector curves indicates a great deal about the operation of a transistor. If the V_{CE} voltage, for example, is held at a constant value of 9 volts, an increase in base current of 5 microamperes to 10 microamperes will cause a change in I_C from 0.5 milliampere to 1 milliampere. The resulting current gain, or beta, produced by this action would be,

$$\begin{aligned}
beta &= \frac{\Delta I_C}{\Delta I_B} \\
&= \frac{0.5\,\text{mA to } 1.0\,\text{mA}}{5\,\mu\text{A to } 10\,\mu\text{A}} \\
&= \frac{500\,\mu\text{A}}{5\,\mu\text{A}} \\
&= 100
\end{aligned}$$

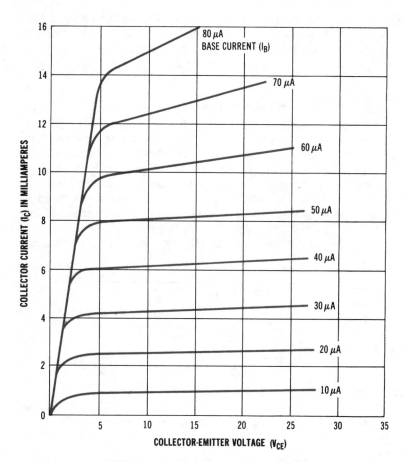

Fig. 4-15. A family of V_{CE}/I_C characteristic curves for a transistor.

The same family of curves also shows that an increase in V_{CE} from 1.0 volt to 9.0 volts, with I_B held at 10 microamperes, will only cause an increase in I_C of 0.5 milliampere to 1.0 milliampere. This indicates that collector current can be more effectively controlled by base current changes than by V_{CE} voltage changes.

When a particular transistor is placed into a circuit with voltage supplied and no signal applied, it is considered to be in a *static* condition of operation. The npn transistor of Fig. 4-16 is used to show this condition. As you will note, this circuit has specific component values assigned. The source voltage is 30 volts, and a 2200-ohm load resistance, R_L, is used. This means that the maximum collector current (I_C) that can occur can be determined by Ohm's law. Mathematically this is expressed as,

$$I_C = \frac{V_{CC}}{R_L}$$
$$= \frac{30 \text{ V}}{2200 \text{ }\Omega}$$
$$= 0.01364 \text{ A}$$
$$= 13.64 \text{ mA}$$

This means that the total range of collector current that can be controlled by the transistor is from 0 milliamperes to 13.64 milliamperes. The actual static condition of operation of this circuit is determined by the selection of a value of base current. In this circuit, the 750-kΩ base resistor determines the operating base current. Mathematically this is determined by dividing the source voltage by the base resistance, or

$$I_B = \frac{V_{CC}}{R_B}$$
$$= \frac{30 \text{ V}}{750 \text{ k}\Omega}$$
$$= 0.00004 \text{ A}$$
$$= 40 \text{ }\mu\text{A}$$

In this case, with 40 microamperes of base current the transistor will only be conducting 6.4 milliamperes of collector current.

In the circuit conditions just described, the base-emitter voltage would be approximately 0.6 volt dc measured across the input, or V_{BE}. The dc voltage measured across the output terminals, which in this case is the collector-emitter voltage, or V_{CE}, would be 15.92 volts and is determined by the collector current, I_C, passing through load resistor R_L. An I_C of 6.4 milliamperes through R_L produces a voltage drop of 14.08 volts across R_L (0.0064 A × 2200 Ω = 14.08 V). Subtracting 14.08 volts from the V_{CC} supply voltage of 30 volts gives us 15.92 volts for V_{CE}. This means that with no

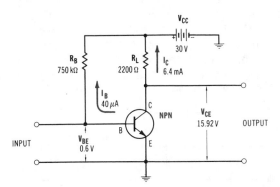

Fig. 4-16. An npn transistor in its static operating condition.

signal applied to the transistor circuit, it will have 15.92 volts dc appearing as output voltage. Since the input voltage of 0.6 volt and the output voltage of 15.92 volts are of a constant value, the amplifier is in a static condition or dc state of operation.

Static Load-Line Analysis

By using the family of V_{CE}–I_C curves of a transistor, its static operating conditions can be determined graphically. Fig. 4-17 shows the collector curves of the transistor being used in the circuit of Fig. 4-16. The diagonal line drawn across the curves is called a *load line*. It represents the extreme conditions of transistor operation for this specific circuit application. In this application, we will assume that the maximum V_{CE} voltage will be

the value of the source, or 30 volts. This point of the load line is located at the 30 V_{CE} mark. This means that when 30 volts of V_{CE} appears across the transistor, there is no I_C current. The cutoff or nonconductive state of operation is represented by this condition.

The other operating extreme occurs when the transistor is conducting the maximum I_C through R_L. We previously determined this value by dividing V_{CC} by the value of R_L (30 V/2200 Ω = 13.64 mA). This point located on the I_C part of the graph indicates the maximum conduction through R_L. When this occurs there will be 0 V_{CE} voltage across the transistor. A straight line connecting these two points together indicates the two extreme operating conditions of the transistor with a 2200-ohm load resistor and a 30-volt source.

Using the load line drawn on the I_C curves of Fig. 4-17, we can now display the static operating conditions of the transistor of Fig. 4-16. The intersection of the 40-microampere I_B curve and the load line is labeled point Q. This base current, you will recall, was established by the value of R_B and V_{CC} previously. Projecting a straight line to the left of the graph from point Q indicates that an I_B of 40 microamperes produces an I_C of 6.4 milliamperes. The resulting V_{CE} that occurs is determined by projecting a line from point Q down to the V_{CE} indicating line. A value of 15.92 volts of V_{CE} is indicated. This means that the voltage across

Fig. 4-17. A family of V_{CE}/I_C characteristic curves with a load line.

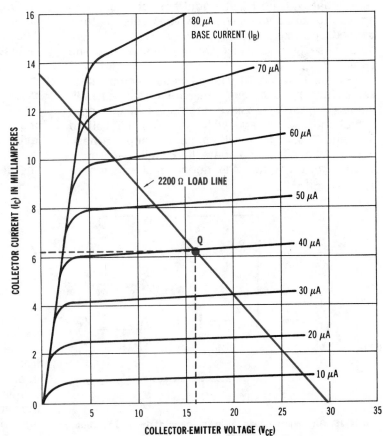

R_L is $V_{CC} - V_{CE}$ or $30 - 15.92 = 14.08$ volts. These values are indicative of the static operating conditions determined previously.

Dynamic Transistor Operation

We will now take a look at the results of applying an ac signal to the base of a transistor in its static state. When this occurs, the transistor has a changing voltage gain or is in a *dynamic* state of operation. Primarily this means that a small ac voltage applied to the input of the transistor causes a variation in the emitter-base voltage (V_{BE}). This change in voltage causes a corresponding change in base current. Variations in base current will, in turn, cause a change in collector current. Ultimately, this action will cause a variation in the I_C passing through R_L which will result in a changing output voltage. Fig. 4-18 shows a representative ac amplifier using the same transistor as that of the static circuit.

To demonstrate dynamic transistor gain, we will first look at the input characteristic of an npn transistor. The input characteristic curve of a typical transistor is shown in Fig. 4-19. Note that the vertical axis of this graph indicates base current (I_B) and the horizontal axis displays emitter-base voltage (V_{BE}). This graph is representative of the input characteristics of the transistor used in Fig. 4-16.

In the static operation of a transistor, you will recall that with no input signal applied there was an emitter-base voltage of 0.6 volt. This, in turn, caused approximately 40 microamperes of I_B. Locate these two points on the input characteristic curve.

Assume now that a 0.2-volt peak-to-peak input signal is applied to our transistor circuit as indicated. During the positive alternation of the input signal the V_{BE} voltage will rise from 0.6 volt to 0.7 volt. This, in turn, will cause a change in base current from 40 microamperes to 60 microamperes. In the same manner, the negative alternation of

the input signal will cause a corresponding reduction in V_{BE} from 0.6 volt to 0.5 volt, and a change in I_B from 40 microamperes to 20 microamperes. Primarily this means that an input voltage change of 0.2 volt peak-to-peak causes a 40-microampere peak-to-peak change in base current.

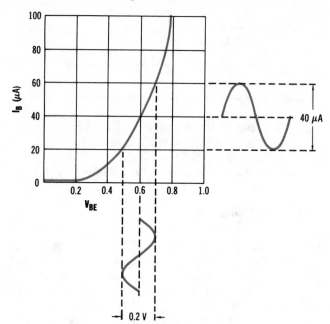

Fig. 4-19. Input characteristic curve for a transistor.

Transferring the changes in base current to the family of I_C curves is shown in Fig. 4-20. The load line drawn on the graph is the same as that achieved with the static circuit of Fig. 4-16. Notice particularly that the total change in I_B as a result of the ac input signal is from 60 microamperes to 20 microamperes. This represents a change in I_B of 40 microamperes. The corresponding change in I_C is from 10 milliamperes to 2.8 milliamperes. This indicates a total change in I_C of 7.2 milliamperes. The current gain, or beta, is, therefore,

$$\text{beta} = \frac{\Delta I_C}{\Delta I_B}$$
$$= \frac{7.2 \text{ mA}}{40 \text{ } \mu\text{A}}$$
$$= \frac{7200 \text{ } \mu\text{A}}{40 \text{ } \mu\text{A}}$$
$$= 180$$

The Greek letter "delta" (Δ) of the formula indicates a change in value.

The resulting output voltage variation of V_{CE} is from approximately 8 volts to 24 volts, or 16 volts peak-to-peak. The dynamic or ac voltage gain (A_V) of this circuit is the output voltage (V_o) divided by the input voltage (V_{in}). Mathematically this is expressed as,

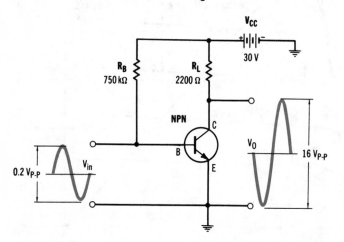

Fig. 4-18. An npn transistor in a dynamic state of operation.

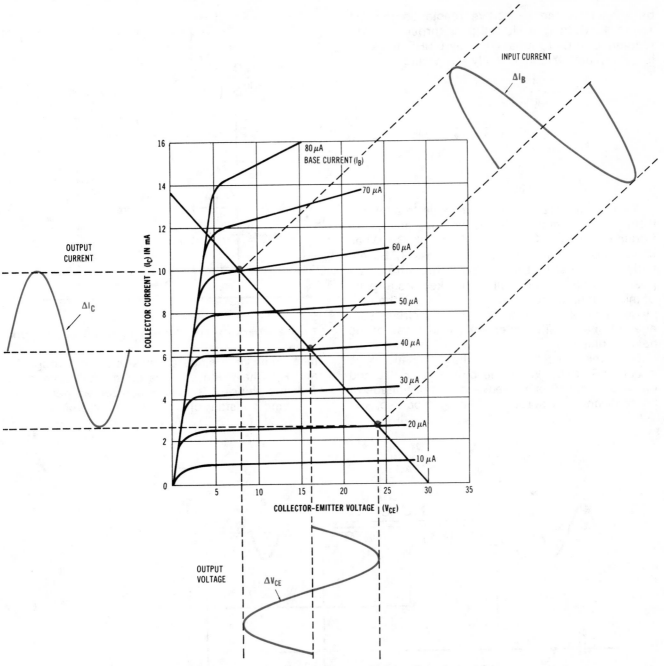

Fig. 4-20. Graphical illustration of the amplifying action of a transistor.

$$A_v = \frac{V_o}{V_{in}} = \frac{\Delta V_{CE}}{\Delta V_{BE}}$$

$$= \frac{16 \ V_{p\text{-}p}}{0.2 \ V_{p\text{-}p}}$$

$$= 80$$

RC Amplifier Coupling

When two or more amplifiers are connected together, the signal being processed must pass from one amplifier to the next without altering the bias voltage of the second amplifier. In ac voltage amplifiers, such as those following the con-

verter, RC coupling is commonly used. Fig. 4-21 shows a single transistor amplifier, similar to the one just discussed, with an added RC coupling network.

The coupling components of the RC network are R_2 and C_c. In order to understand the operation of this coupling network, we shall discuss it first from a capacitive reactance point of view and second in terms of a time constant. The capacitive reactance (X_c) of C_c is infinite when the voltage across it is of a constant value or dc. We therefore say that it blocks dc or responds as an open circuit to direct current. Since this is true, as can be seen from the capacitive reactance formula, $X_c = 1/2\pi fC$, then no dc will pass through R_2. The reactance of C_C to an ac such as 60 Hz, however, is quite small. Mathematically this would be,

$$X_c = \frac{1}{2\pi fC}$$

$$= \frac{1}{6.28 \times 60 \times 10 \times 10^{-6}}$$

$$= \frac{1}{3.768} \times 10^3$$

$$= 265.39 \ \Omega$$

A capacitive reactance of this value to ac is obviously quite small. This means that ac of this frequency will pass very easily through C_C. When this occurs, ac will be developed across R_2. Essentially this means that ac will pass through C_C, appear across R_2, and dc will be blocked. As a result of this action, ac signals can be passed to the next stage of amplification without the dc circuit voltages of one amplifier interfering with those of the next amplifier.

The time constant (T) of the RC network is equal to $R_2 \times C_C$. When the circuit is first turned on, and no signal is applied, a collector voltage of 30 volts appears across C_C. This, of course, will

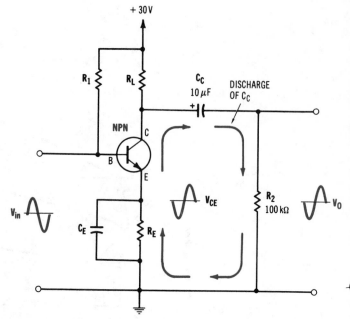

Fig. 4-21. Transistor amplifier with an RC coupling network.

take some time since C_C must charge through R_2 and R_L. The value of R_L is quite small compared with the value of R_2 and is considered negligible. The time constant of C_C and R_2 is, therefore,

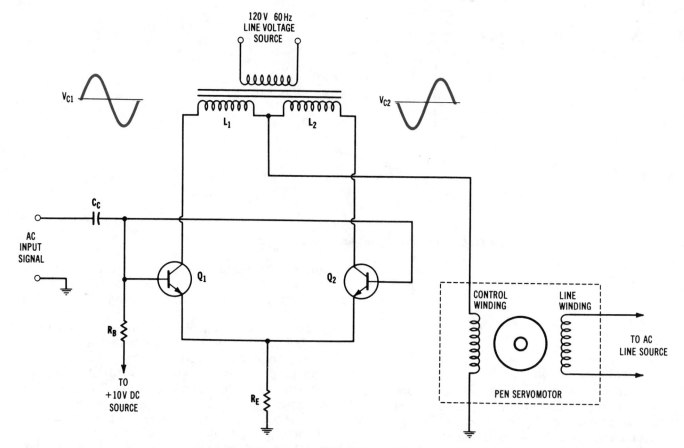

Fig. 4-22. Power-amplifier motor-control circuit.

$$T = R_2 \times C_C$$
$$= 100\,k\Omega \times 10\,\mu F$$
$$= 1 \times 10^5 \times 10 \times 10^{-6}$$
$$= 10^6 \times 10^{-6}$$
$$= 1\ \text{second}$$

Complete charge of C_C will thus occur in 5 seconds.

When an ac signal is applied to the base of Q_1, it will cause an amplified ac signal to appear at the collector as indicated. The positive alternation of the input signal will produce a corresponding decrease in V_{CE} voltage. With C_C charged to the $+V_{CC}$ voltage, it will attempt to discharge through R_2, to ground, and through Q_1 as indicated by the arrows. Neglecting the internal resistance of Q_1, it would require at least 1 second for C_C to discharge. With the total time of the positive alternation only half of 1/60, or 1/120 of a second, C_C does not have adequate time to discharge. The amount of discharge is actually equivalent to the amount of voltage drop due to X_C. For all practical purposes, C_C remains charged to nearly the value of V_{CC}.

During the negative alternation of the input signal, the V_{CE} voltage rises in value. When this occurs, C_C attempts to charge to the increased V_{CE}

voltage through R_2. In the short time that this voltage is up in value, C_C charges and a current conducts through R_2.

The charge and discharge action of C_C causes a corresponding current conduction through R_2. In effect, this means that an ac voltage appears across R_2. This voltage is representative of the V_{CE} signal of the transistor, and is passed through C_C. The input of a second stage of amplification connected across R_2 would, therefore, receive the resulting ac signal through RC coupling. This signal is an excellent reproduction of the original input signal, with suitable gain, and it is inverted 180°. Today RC coupling is commonly used in electronic recording instruments.

POWER AMPLIFIER (MOTOR-CONTROL CIRCUIT)

The last stage of amplification in an electronic recording instrument is primarily responsible for manipulation of the recording pen motor. This amplifier, depending on the design of the recorder, drives the motor in such a way that it causes the pen to move on a chart according to variations in the input signal. As a general rule, a feedback signal from the pen location is returned to be compared with the measured input signal.

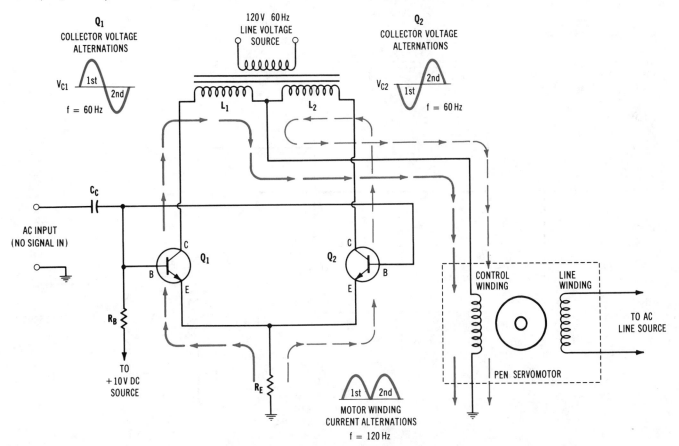

Fig. 4-23. Operation of motor-control circuit with no ac input signal.

Typically, one stage of power amplification is used in the motor-control amplifier. In the discussion of power amplifiers that follows, we will use two power transistors in a parallel configuration. Each transistor controls half of the resulting output signal.

Fig. 4-22 shows a representative motor-control power-amplifier circuit that is used in several different recorders today. An RC network is used to couple the voltage amplifier output signal to the bases of both transistors. An emitter resistor, R_E, is commonly connected to both transistor emitters. Collector voltage for each transistor is supplied through a special secondary winding of the power transformer. The winding is center tapped with equal amounts of voltage supplied to each transistor. These voltages are of equal amplitude and are 180° out of phase.

Amplifier Operation (No Signal Input)

No signal input to the power amplifier is provided when the measuring circuit and pen location feedback signal are properly balanced. Fig. 4-23 shows a schematic of the motor-control circuit in this condition of operation. With no signal applied, the bases of both transistors do not receive

a signal. They are, however, forward biased by R_B and in a static condition of operation. With ac applied to the collector of each transistor, it means that the transistors will be alternately conductive. For example, during the first alternation of the transformer input, the collector of Q_1 is positive, while the collector of Q_2 is negative. You will recall that conduction of a transistor only occurs when the collector-base junction is reverse biased. This means that Q_1 is conductive and Q_2 is nonconductive during this alternation of the ac collector voltage. Q_2 is not conductive because its collector-base junction is improperly biased. As you will note, current, indicated by the solid arrows, passes through R_E, Q_1, L_1, the servomotor winding, and returns to ground.

During the next alternation, the conduction of Q_1 and Q_2 will switch. Q_2 now becomes conductive because its collector is positive. This current is indicated by the broken arrows. The path is through R_E, Q_2, L_2, the servomotor winding, and ground. Since the resulting current through both transistors is in the same direction, the voltage drop across the motor winding has the same polarity for each half-cycle of the input. Two resulting pulses of motor winding current are produced for

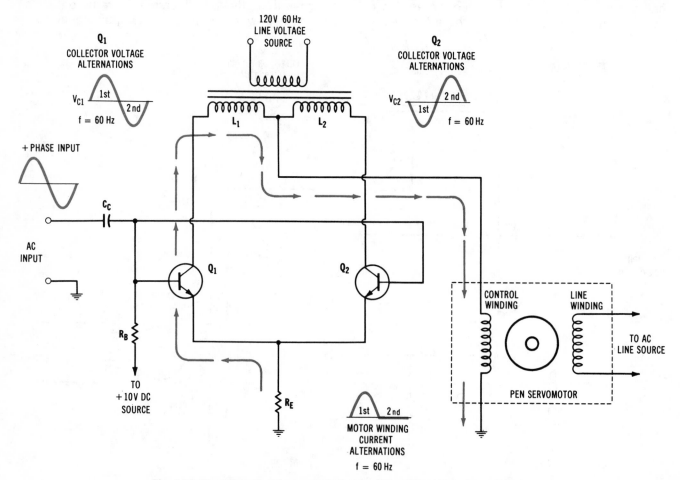

Fig. 4-24. Operation of motor-control circuit with positive phase input signal.

76

each cycle of the applied collector voltage input. The frequency of these pulses is 120 Hz or twice the 60-Hz line voltage. The servomotor, therefore, remains in a stationary position. Ideally this represents a condition of balance. The pen should be at its center resting position, or at its zero reference location.

Amplifier Operation (With Signal Input)

Assume now that the circuit of Fig. 4-24 has an ac signal applied to its input. This condition would take place when an input signal is applied to the recorder input. When this occurs, the pen location and input signal are out of balance and an error signal is generated, converted to ac, amplified, and applied to the motor-control power amplifiers.

With the bases of power amplifiers Q_1 and Q_2 commonly connected together, any input signal will appear the same at both bases. The signal of V_{C1} will, for example, appear as indicated in Fig. 4-24. For the positive input alternation, note that Q_1 is conductive and that Q_2 is not conductive. This is primarily due to the polarity of the collector voltage and the base voltage. The base of Q_1 is forward biased and the collector is reverse biased. At the same time, Q_2 is nonconductive be-

cause, although its base is forward biased, its collector is improperly biased because of the polarity of the ac line voltage. The resulting current path for this alternation is, therefore, through R_E, Q_1, L_1, the servomotor winding, and ground.

For the next alternation of the input signal, the bases of both Q_1 and Q_2 will swing negative. This, of course, reverse biases the bases of both transistors regardless of the polarity of the collector voltages. In this condition, no resulting current conducts through either transistor. An output, as indicated, only occurs during the positive alternation of the base input signal. The frequency of the output is 60 Hz.

If the polarity of the input signal is reversed, as indicated in Fig. 4-25, so that the first alternation is negative, the resulting output will shift. In this case, the bases of both Q_1 and Q_2 will be reverse biased because of the negative alternation. No current will conduct through either transistor in this condition of operation regardless of the collector polarity. During the next alternation of the input, which is positive, the bases of both Q_1 and Q_2 will be forward biased. The collector of Q_1 will now have the wrong polarity while the collector of Q_2 will have the correct polarity. Therefore, con-

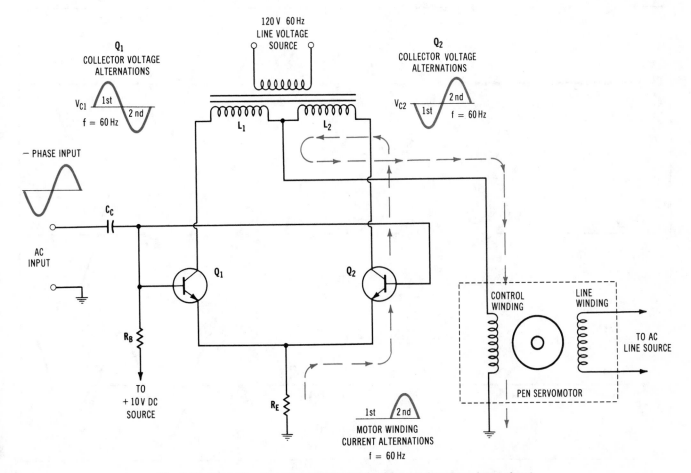

Fig. 4-25. Operation of motor-control circuit with negative phase input signal.

duction will occur through R_E, Q_2, L_2, the motor winding, and ground. The resulting output of this sine-wave input is one pulse of output. This occurs at a frequency of 60 Hz and is shifted 180° with respect to the previous input signal.

The input signal to the power amplifier may be in phase with the line voltage supplied to the transformer, or it may be 180° out of phase with it. This is essentially determined by the direction of unbalance in the measuring circuits.

POWER AMPLIFIER OUTPUT EFFECT ON THE BALANCING MOTOR

In this section the reactions of the balancing motor due to amplifier unit outputs will be discussed. The balancing motor itself will be treated as a "black box." Since most instruments use similar balancing motors, the split-phase motor will be dealt with as a separate unit.

Fig. 4-26 shows the phase relationship of the amplifier unit inputs, outputs, and the line phase. The amplifier output is the current conducting through one phase winding of the balancing motor. The relationship between the currents in the windings of the motor determines the torque produced

by the motor. A phase shift of about 90° is necessary for motor rotation. The current through the winding supplied by the power supply is always 90° out of phase with the line voltage. The inputs of the amplifier units are developed by the converter circuits.

The power amplifier outputs are not pure sine waves. They are a mixture of a dc component, a 60-Hz ac component, and higher frequency components called harmonics. The only part of this output that will produce motor rotation is the 60-Hz component. When the input is zero, there is no motor rotation. The amplifier output has a frequency of 120 Hz. There is no 60-Hz ac component. Therefore, no motor rotation results. You will notice that the output developed from V_1 does have a 60-Hz ac component. This 60-Hz ac voltage is shown in Fig. 4-26. This voltage leads the line phase by 90°. Rotation of the motor occurs to rebalance or null the measuring circuits. The output resulting from V_2 is 180° out of phase with the voltage just described. The phase shift was determined by the measuring circuits. Its ac component lags the line phase by 90°. This is the condition necessary to produce rotation of the motor in the reverse direction.

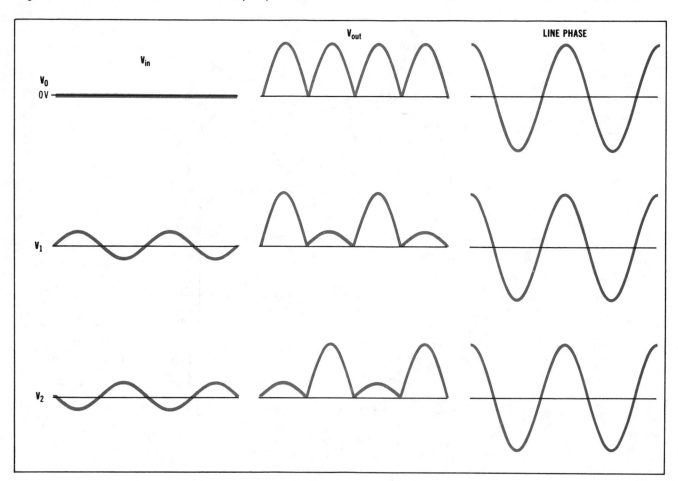

Fig. 4-26. Phase relationships in amplifier motor-control circuit.

We will assume that the input voltage of Fig. 4-24 is in phase with the line voltage. The resulting output will occur during the first alternation. In Fig. 4-25 the reverse is true. The input signal is 180° out of phase with the line voltage. The resulting current output, therefore, occurs during the second alternation of the input signal. This means that the output pulse will shift 180° according to the phase polarity of the input signal. The shifting of this signal with respect to the voltage applied to the line winding ultimately determines the direction of motor rotation.

SUMMARY

Electronic recording instruments in general are designed to provide a type of graphic display of a particular quantity being measured. Instruments of this type record on strip charts, circular charts, or by hand deflection on a meter scale.

The major parts of a recorder are the null balancing mechanism, power supply, amplifiers, converters, power amplifier, servomotor, and recording mechanism.

The power source of a recorder is designed to supply the operating voltages needed to make the recorder function. Rectification, filtering, and regulation are common parts of the power supply.

Dc to ac converters are a common part of the electronic recorder. This circuit simply changes dc into ac with a vibrating reed device known as a chopper.

A number of voltage amplifiers are needed to increase the amplitude level of the signal to a value that will drive the power amplifier. Transistors are commonly used today. They are designed to have ac gain by plotting a load line on a family of IC curves. Beta is an expression of transistor gain. The input amplifiers of a recorder are small-signal voltage amplifiers. The final output signal of these amplifiers is used to drive the power amplifier.

Power amplifiers in a recorder are commonly used to drive the servomotor mechanism. The mechanism moves the pen on a chart and actuates the balancing mechanism. The servomotor moves the pen and balancing mechanism according to the phase of the measured ac signal compared to the ac line voltage reference source.

CHAPTER 5

Honeywell ElectroniK Recorders

INTRODUCTION

There are numerous commercial electronic recording instruments available today for use in industrial applications. As a general rule, each manufacturer tries to do something unique with its instrument to improve performance or make it more competitive than the others. Such things as multipoint recorders, different recording techniques, physical size, and electronic circuitry account for some of the differences in these instruments.

A person working with electronic recording instruments should have some understanding of the circuitry needed to make the instrument perform.

Due to the number of instruments available, it is difficult to single out one particular instrument and refer to it as an industrial standard. The Honeywell ElectroniK 15 recorder discussed in this chapter is only one example of a number of instruments available today. The circuits discussed apply specifically to the Honeywell instrument. The operation of this instrument could, however, apply to a number of other commercial instruments. When working with a specific recorder, the manufacturer's operational and instructional literature should always serve as the primary reference source.

The Honeywell ElectroniK 15 universal multipoint recorder is shown in Fig. 5-1. This particular

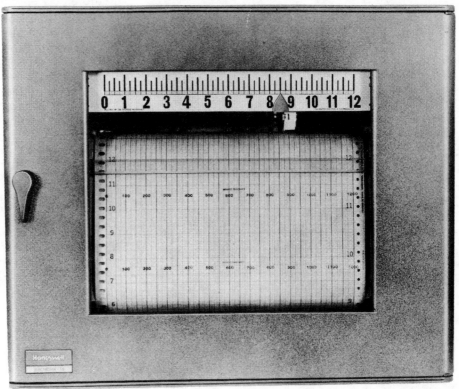

Fig. 5-1. The Honeywell ElectroniK 15 universal multipoint recorder.

recorder has the capability of being changed so that it can record information from 2 to 24 points. Converting the number of points being measured can be done within a matter of seconds by altering a shorting plug. This type of recorder uses a syncro-balancing motor-control unit.

FUNCTIONAL UNITS

The functional units of the ElectroniK 15 recorder are shown in block-diagram form in Fig. 5-2. A recorder of this type contains two major divisions with several discrete electronic functions in each division. The amplifier unit contains a power supply, filter, chopper, voltage amplifier, and a driver stage. The motor unit contains a power amplifier and a balancing motor.

The operation of each discrete functional block of the recorder will be briefly discussed in this chapter. The components making up each block involve the principles of resistance, capacitance, and inductance, as well as diode and transistor operation. These components and their principles of operation will serve as the basis for the discussion. In a sense, we will be using the recorder as a means of discussing specific component applications. Through this approach, you will be able to increase significantly your understanding of electronic theory.

GENERAL RECORDER OPERATION

Using the block diagram of Fig. 5-2 as a reference, we will first discuss the recorder in general operational terms. Initially, the input to the amplifier is a dc signal that varies according to the measured value of the variable under consideration. It may be of either positive or negative polarity. This signal then passes through the filter, which removes extraneous noise. The chopper then changes dc into a series of pulses that have a character-

istic ac shape. This signal is amplified and ultimately applied to the driver stage. It is then added to the rectified sine wave that drives the motor unit. Motor rotational direction is based on the phase difference between this ac signal and a fixed phase signal from the power line. A positive dc input causes clockwise motor rotation until balance is achieved, and a negative input causes counterclockwise rotation. The "dither" signal applied to the input is a small voltage that causes a slight oscillation in motor position when it reaches a state of balance.

Power Supply

The power supply of the Honeywell ElectroniK recorder provides ac to the motor as a line reference source and also develops dc voltage for component operation. A simplified schematic of the power supply is shown in Fig. 5-3. The main rectifier employing CR_5 and CR_6 is a full-wave circuit with two zener diode regulators. Three different dc voltage values are developed for circuit operation. Capacitor C_{13} is part of the RC filter. Diode CR_3 is added to the circuit to prevent reverse voltage spikes from entering the dc supply circuit. It is forward biased, which permits full conduction capabilities during normal operation.

Diode CR_1 and its associated components serve as a half-wave rectifier to develop a negative dc voltage. This voltage is regulated by VR_3 and filtered by C_2, R_6, and R_2. Note that CR_1 is placed in the power supply in a reverse polarity. The negative dc voltage developed by this rectifier is ultimately used to bias the collector of the waveshaping amplifier of the chopper circuit.

Input Filter

The input filter of the Honeywell recorder is shown in Fig. 5-4. A three-section RC network is formed by R_9-C_4, $R_{10}-C_5$, and $R_{12}-C_6$. A dc signal applied to the input charges each section of the

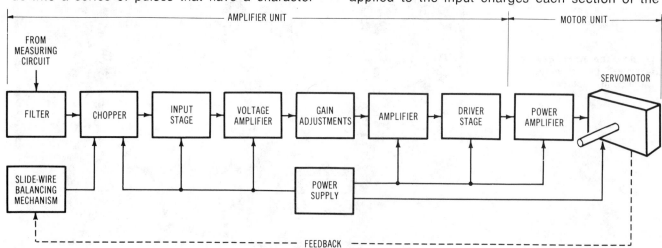

Fig. 5-2. Block diagram showing the functional units of the Honeywell ElectroniK 15 recorder.

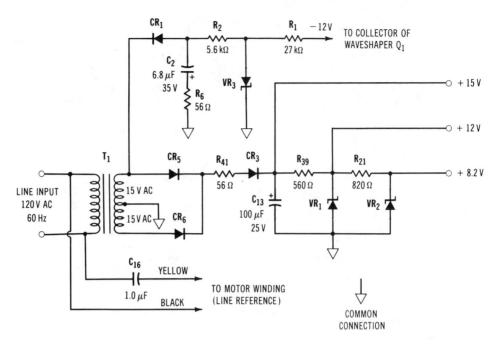

Fig. 5-3. Simplified schematic of the power supply used in the Honeywell ElectroniK 15 recorder.

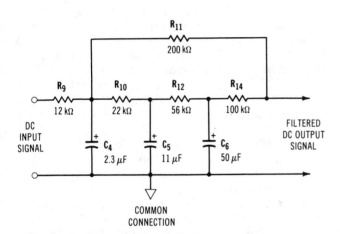

Fig. 5-4. The input filter of the Honeywell ElectroniK 15 recorder.

filter in succession. Through this action, the filter removes any noise or extraneous ac voltages that may appear at the input. Resistor R_{11} is used to couple the input to the output so that any ac passing through the filter will be of equal amplitude but 180° out of phase. As a result of this action, any resulting ac is canceled by phase inversion. A discussion of phase shifting and component influence on ac phase will occur later in this chapter.

Chopper/Waveshaper

The chopper of the Honeywell recorder is an electronic circuit that changes dc to ac. This circuit supersedes the mechanical vibrating chopper discussed in Chapter 4.

A simplified schematic of the electronic chopper is shown in Fig. 5-5. As you will note, this circuit employs an insulated-gate field-effect tran-

sistor (IGFET) and a bipolar transistor. Q_1 is a waveshaping transistor that changes the ac secondary voltage of transformer T_1 into a negative-going square wave. The base circuit of Q_1 is preceded by an RC low-pass filter consisting of R_4, R_5, and C_3. This filter rejects high frequency by sending it into ground and passes low frequency such as 60 Hz. As a result of this, 60 Hz from T_1 is applied to the base of Q_1.

Diode CR_1 is likewise connected to the secondary voltage of T_1. In this case, CR_1 is a half-wave rectifier with an RC filter. C_2, R_6, and R_2 are the filter components. Zener diode VR_3 regulates the dc voltage to −12 volts dc. This voltage is used to reverse bias the collector of Q_1. As a result of this circuit, Q_1 produces a square-wave output that changes between +6 and −6 volts.

The IGFET, labeled Q_2 in the diagram, has three signals applied to it. The square-wave signal from Q_1 is supplied to the gate through capacitor C_1. This signal causes the source-drain resistance of Q_2 to vary between a very high value to a few hundred ohms on alternate halves of the waveform. This, in turn, causes the output of Q_2 to alternately respond to the dc slide-wire feedback signal at the source and the dc input signal at the drain. The output of Q_2, therefore, corresponds to these two input signal levels. Input and feedback signals of equal value will result in zero ac output. An imbalance in the two input signals will result in an ac output signal that will shift phase in either the positive or negative direction. Through this circuit, the dc input will produce an ac output that shifts phase according to the polarity of the input voltage.

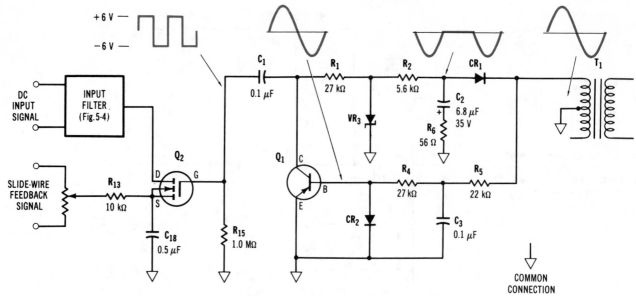

Fig. 5-5. Chopper/waveshaper circuit of the Honeywell ElectroniK 15 recorder.

Voltage Amplifiers

The next function of the ElectroniK 15 recorder is voltage amplification. Fig. 5-6 shows a simplified circuit of the input voltage amplifier following the chopper. Note that Q_3 is a junction field-effect transistor, and that Q_4 and Q_5 are bipolar transistors.

Transistor Q_3 is a high-impedance n-channel JFET. The ac signal from the chopper is applied to its input through capacitor C_7. The gate of a JFET, you will recall, varies conduction through the channel. It is normally reverse biased. The applied ac signal will add to or reduce the effect of reverse biasing. As a result of this, the ac signal controls the source-drain current passing through the channel. In effect, a small change in

reverse bias voltage will have a rather significant influence on the drain current passing through R_{17}. The resulting voltage drop is representative of the amplified output signal. This is then coupled to Q_4 through C_8. Resistor R_{44} is a source biasing resistor which stabilizes amplification. The gate resistor, R_{16}, serves as a discharge or return path for the ac signal passing through C_7. This amplifier has a characteristically high input impedance and a rather low output impedance.

The output signal of Q_3 is applied to the input of Q_4 through capacitor C_8. Transistor Q_4 is forward biased by a voltage divider composed of R_{18}, R_{20}, R_{23}, R_{24}, and CR_4. Q_4 is connected in an emitter-follower circuit configuration. It has high input impedance and low output impedance. The output signal, which has a gain of slightly less than one,

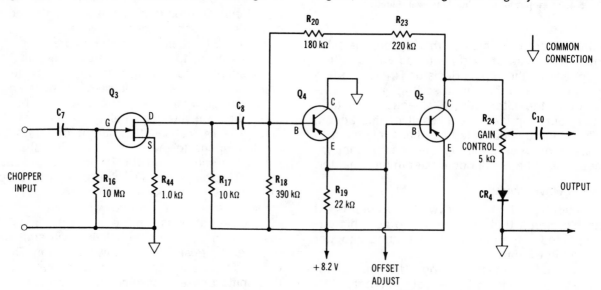

Fig. 5-6. Input voltage amplifier stage of the Honeywell ElectroniK 15 recorder.

is developed across the emitter resistor, R_{19}. In this situation, Q_4 is used as an impedance matching device to develop maximum signal transfer to Q_5.

The emitter output of Q_4 is coupled directly to the base of Q_5. Base biasing is achieved by R_{19} and the offset adjust control (not shown). This transistor has a low input impedance. It is, however, connected in a common-emitter circuit configuration with the output developed across the collector resistor, R_{24}. This resistor also serves as the gain adjusting control. The output of the amplifier, which is passed through capacitor C_{10}, can be adjusted to any level of gain depending on the desired level of sensitivity. The current gain, or beta, of the composite input amplifier can, therefore, be adjusted to meet the application needs of the input signal. CR_4 is called a "pedestal-diode," which is used to prevent the polarity of C_{10} from developing a reverse voltage for the next amplifier during low gain settings.

Direct-Coupled Amplifiers

Transistors Q_6, Q_7, and Q_8 serve as a second block of voltage amplifiers. As you will note in Fig. 5-7, these transistors are all bipolar transistors connected in a direct-coupled circuit configuration. The input to this circuit is through C_{10}. Transistor Q_6 is biased by emitter resistor R_{27}. Resistors R_{25} and R_{26} serve as a discharge path for C_{10}. The ac signal appearing across R_{26} is applied to the base of Q_6. Resistor R_{28} serves jointly as a load resistor for Q_6 and as a base bias resistor for Q_7. This positive voltage reverse biases the collector of Q_6 and forward biases the base of Q_7 at the same time. Any variation in the

collector current of Q_6 is applied directly to the base of Q_7.

The output of Q_7 is coupled directly to the base of Q_8 in the same manner that Q_6 and Q_7 were coupled. The emitters of these two transistors are both connected to the common or negative side of the power supply. The composite output of the three transistors appears across C_{14}. A rather substantial amount of signal gain is achieved through this block of amplifiers.

Driver Amplifier/Power Amplifier

The driver amplifier, Q_9, of Fig. 5-8 employs RC coupling between the collector of Q_8 and its base. Capacitor C_{14} serves as the coupling capacitor, and resistor R_{33} serves as the discharge resistor. Voltage changes at the top of R_{33} are applied to the base of Q_9 through R_{45}. A voltage divider made up of R_{40} and R_{33} establishes forward bias voltage for the base of Q_9. The emitter resistor, R_{42}, determines the transistor operating point and establishes the gain capabilities of the driver.

The output of the driver amplifier is directly coupled to the base of the power amplifier, Q_{10}. The resistance of Q_{10} serves as the load resistor for Q_9. Variations in the collector current of Q_9 and the base of Q_{10} are needed to control the high current passing through Q_{10}. The total current of the power amplifier is ultimately determined by the gain range switch. The position of this switch determines the value of the emitter resistor.

The dc voltage applied to the collector of Q_{10} is the unfiltered full-wave output of the power supply. This voltage provides the conventional reverse bias for the collector. With the ac signal applied to the input taking on either a positive or nega-

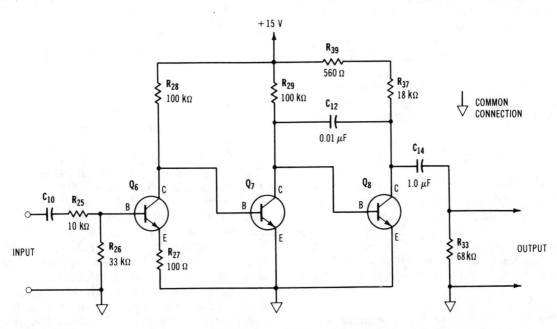

Fig. 5-7. Direct-coupled voltage amplifiers used in the Honeywell ElectroniK 15 recorder.

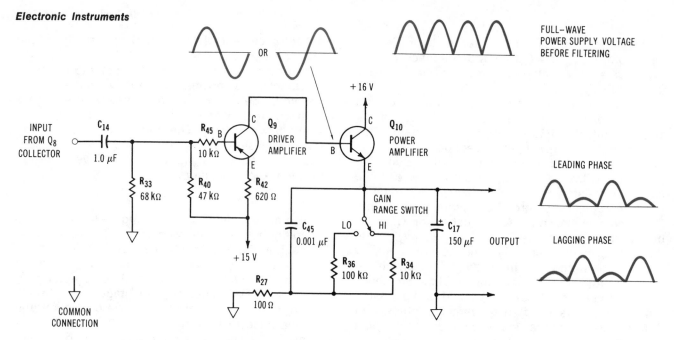

Fig. 5-8. Driver/power amplifier stage used in the Honeywell ElectroniK 15 recorder.

tive phase, as indicated, the output will have either a leading or lagging output across C_{17}. If the first alternation of the input is positive, the output will produce positive pulses or be of a leading phase as indicated. If the first alternation is negative, the output pulse will have reduced amplitude as indicated. Through this mixing process, the output has a phase shifting capability that is dependent upon the polarity of the original dc input signal.

The power amplifier of this unit is considered to be single ended. The voltage gain produced by Q_{10} is less than one, but is still great enough to alter the control winding voltage of the servomotor. Typical voltage values are 40 volts peak-to-peak due to the previous level of amplification. The voltage output of the power amplifier is considered to be control voltage when it is applied to the servomotor. The resulting operation of the power amplifier with respect to the full-wave rectifier input is called *null detection.* This effect is similar to the two-amplifier null detector of Chapter 4.

PHASING

From the preceding discussions, it can be seen that phasing is of importance in the operation of recording instruments. Any discussion of phase relationships applies, in many cases, to all instruments since they are very much alike in circuitry. Some background information on the effect of components on phase will be necessary.

Resistive AC Circuit

Resistance is defined as that component that causes a voltage drop that is directly proportional

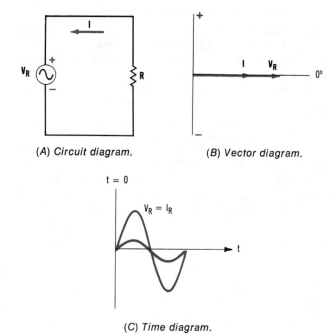

(A) Circuit diagram.

(B) Vector diagram.

(C) Time diagram.

Fig. 5-9. Purely resistive ac circuit.

to the current through it. This relationship (Ohm's law) holds true for current that varies with time as well as for constant current. Fig. 5-9 shows this relationship in the resistive circuit. The current and voltage in this circuit are obviously in phase as can be seen in the time diagram of Fig. 5-9C. A vector diagram for this circuit is also shown in Fig. 5-9B. A vector is a quantity that indicates not only magnitude but also direction. The length of the vector will represent the magnitude of either a current or a voltage; the direction of a vector will represent the phase relationships of currents and voltages. Two vectors pointing in the same

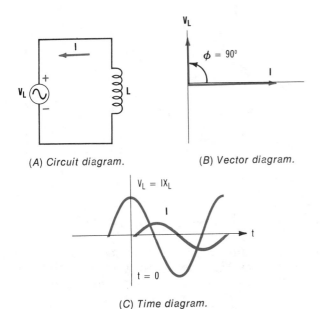

(A) Circuit diagram.

(B) Vector diagram.

(C) Time diagram.

Fig. 5-10. Purely inductive ac circuit.

direction, as in Fig. 5-9B, are in phase. Vectors are merely a method of visually representing voltages and currents and their phase relationships. A vector that is rotated counterclockwise from a reference vector is said to have a leading (positive) phase angle. The amount by which it leads the reference is indicated by the amount of rotation. A vector that is rotated clockwise from a reference vector has a lagging (negative) phase angle.

Inductive AC Circuit

Inductance is defined as that circuit element that has a voltage proportional to the rate of change of current through it. Simply stated this means that an inductor opposes changes in current. Since voltage is proportional to the rate of change of

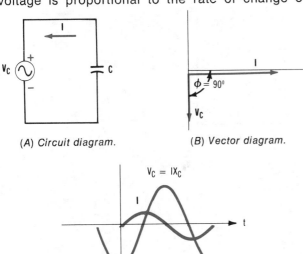

(A) Circuit diagram.

(B) Vector diagram.

(C) Time diagram.

Fig. 5-11. Purely capacitive ac circuit.

current, it is out of phase with the current, as indicated in Fig. 5-10C. The ratio between the voltage across an inductor and the current through it is called inductive reactance ($V_L = IX_L$). Reactance implies a 90° phase shift as opposed to resistance. The vector diagram (Fig. 5-10B) indicates that voltage leads current through an inductor by 90°.

The phase shift in this circuit results from the fact that the magnetic field produced by a coil varies with the current. A changing magnetic field in turn induces a counterelectromotive force into the circuit. The voltage drop, therefore, is produced by the change in current and not by its magnitude.

Capacitive AC Circuit

Capacitance is defined as that circuit element that has a current proportional to the rate of change of voltage across it. Simply stated this means that a capacitor opposes changes in voltage. Since current is proportional to the rate of change of voltage, it is out of phase with the voltage as indicated in Fig. 5-11C. The ratio between the voltage and current of a capacitor is called capacitive reactance ($V_C = IX_C$). Both inductive and capacitive reactance have the dimensions of ohms.

The vector diagram (Fig. 5-11B) indicates that capacitive current leads the voltage by 90°. Conversely, voltage lags the current. The phase shift in this circuit results from the ability of the capacitor to store energy in an electric field; i.e., its ability to charge. As described in the chapter concerning time constants, any increase in the voltage across the capacitor results in a decrease in current, thereby decreasing its charging rate.

Series RL Circuit

Fig. 5-12A shows a series RL circuit. Actually, any circuit containing an inductor involves some resistance because the wire used to wind a coil has some resistance. A fixed resistive component may also be inserted into the inductive circuit. A series circuit is defined as a circuit that provides a single path for current. Since this is true, the same current will pass through each element in the series circuit. Current, therefore, is used as the reference in the series circuit since it is the common value for each component.

If the effective current is measured in this circuit, then the voltage across R, (V_R) and the voltage across L (V_L) can be computed. By Ohm's law, $V_R = IR$ and $V_L = IX_L$. As indicated in the time diagram for this circuit (Fig. 5-12C), V_R is in phase with the current, while V_L leads the current by 90°. Kirchhoff's voltage law states that in a series dc circuit, the sum of the voltage drops in the circuit equals the applied voltage (V_A). This is also true in an ac circuit at any instant of time. It is obvi-

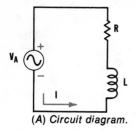

(A) Circuit diagram.

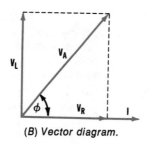

(B) Vector diagram.

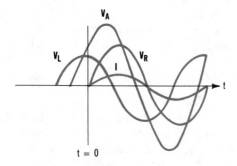

(C) Time diagram.

Fig. 5-12. Series RL circuit.

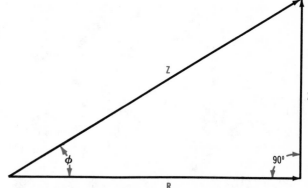

Fig. 5-13. Impedance vector diagram.

ous then that peak or effective values cannot be added directly since they do not occur at the same time. In the time diagram (Fig. 5-12C), instantaneous values of voltage have been added to produce the sine wave that represents the applied voltage (V_A). This is a most laborious process. The same result is obtained by adding V_R and V_L vectorially. The vector sum of two vectors is the diagonal of the parallelogram whose sides are composed of the vectors V_R and V_L. The phase angle (ϕ) between I and V_A is also obtained from the vector diagram (Fig. 5-12B). Mathematically, V_A can be found by the Pythagorean theorem since V_R and V_L form the legs of a right triangle, and V_A is its hypotenuse. Therefore,

$$V_A = \sqrt{V_R{}^2 + V_L{}^2}$$

and

$$\phi = \text{arc tan } \frac{V_L}{V_R}$$

By convention, reference vector I is drawn horizontally from left to right and is given a phase angle of 0°.

Resistance and inductive reactance are directly proportional to V_R and V_L in the series circuit. Since this is true, a vector diagram representing reactance and resistance can be drawn (Fig. 5-13). The vector sum of resistance and reactance in a series circuit is referred to as the *impedance* (Z) of the circuit.

The impedance value is the ratio of the applied voltage and the current ($V_A = IZ$). The phase angle (ϕ) is the same as that found in the voltage diagram since these two triangles are similar; i.e.,

their sides are proportional. Impedance could be described as the total opposition to ac current. Its symbol is Z, and the unit of measure is the ohm.

Series RC Circuit

Fig. 5-14 indicates the voltage and current relationships found in an RC circuit. We now know that the voltage across the capacitor lags the current by 90°. We can see from the vector diagram (Fig. 5-14B) that V_A also lags the current. The size of the phase angle (ϕ) depends on the relative magnitudes of V_R and V_C. The only difference between the capacitive circuit and the inductive circuit is that in the capacitive circuit current leads voltage, while in the inductive circuit, current lags voltage.

Solving a sample problem should help us to see the applications of the principles discussed. Fig. 5-15A indicates the known values in the circuit. The resistance, R, the capacitance, C, the current in the circuit, and the frequency of the applied voltage are the known quantities. Since frequency

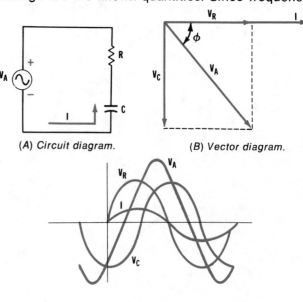

(A) Circuit diagram.　　*(B) Vector diagram.*

(C) Time diagram.

Fig. 5-14. Series RC circuit.

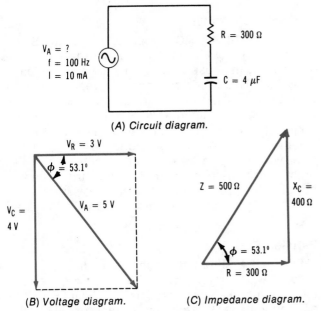

(A) Circuit diagram.

(B) Voltage diagram.

(C) Impedance diagram.

Fig. 5-15. Series RC circuit solved by vectors.

and capacitance are known, the capacitive reactance can be computed as follows:

$$X_C = \frac{1}{2\pi fC}$$

$$= \frac{1}{6.28 \times 100 \times 4 \times 10^{-6}}$$

$$= \frac{0.159}{4} \times 10^4$$

$$= 0.04 \times 10^4$$

$$= 400 \ \Omega$$

Also,

$$V_R = IR$$

$$= 10 \times 10^{-3} \times 3 \times 10^2$$

$$= 3 \text{ volts}$$

$$V_C = IX_C$$

$$= 10 \times 10^{-3} \times 4 \times 10^2$$

$$= 4 \text{ volts}$$

From the voltage vector diagram (Fig. 5-15B), the applied voltage, V_A, and the phase angle, ϕ, could be obtained graphically. Mathematically,

$$V_A = \sqrt{V_R{}^2 + V_C{}^2}$$

$$= \sqrt{3^2 + 4^2}$$

$$= \sqrt{25}$$

$$= 5 \text{ volts}$$

$$\phi = \text{arc tan} \frac{V_C}{V_R}$$

$$= \text{arc tan} \frac{4}{3}$$

$$= \text{arc tan } 1.333$$

$$= 53.1°$$

Following the same procedure with the impedance diagram (Fig. 5-15C),

$$Z = \sqrt{R^2 + X_C{}^2}$$

$$= \sqrt{300^2 + 400^2}$$

$$= \sqrt{25 \times 10^4}$$

$$= 500 \ \Omega$$

$$\phi = \text{arc tan} \frac{X_C}{R}$$

$$= \text{arc tan} \frac{400}{300}$$

$$= \text{arc tan } 1.333$$

$$= 53.1°$$

This agrees with the impedance obtained by Ohm's law,

$$Z = \frac{V_A}{I}$$

$$= \frac{5 \text{ V } \angle 53.1°}{10 \text{ mA}}$$

$$= 500 \ \Omega \text{ at } 53.1°$$

Series RLC Circuit

Fig. 5-16A shows a series RLC circuit. Current is still the reference vector since it is common to all components. V_R is in phase with I. V_L leads the current by 90°. To find V_A, these three vectors must be added vectorially. Any two of these vectors are added first and the sum added to the remaining vector. In the example shown in Fig. 5-16B, V_L and V_R were added to produce $V_L + V_R$. This vector was then added to V_C to produce V_A. Since V_L and V_C are in opposite directions, their sum is merely the difference in the magnitudes; i.e., $V_L - V_C$. Mathematically, $V_A = \sqrt{V_R{}^2 + (V_L - V_C)^2}$. You will notice that if X_L and X_C are equal, V_A is equal to V_R. Under these conditions, the circuit is said to be resonant. This will happen at one frequency for every capacitor-inductor combination. At this frequency, you will also notice that $Z = R$.

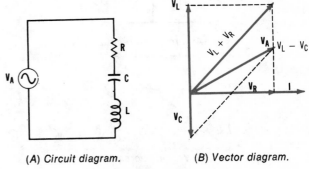

(A) Circuit diagram.   (B) Vector diagram.

Fig. 5-16. Series RLC circuit.

Impedance at this frequency, therefore, is at its minimum. If it were possible for a circuit to have a resistance of 0 ohms, it would occur at resonance. If Z is at a minimum, then current is maximum in a series circuit at resonance.

Parallel RL Circuit

In a parallel circuit, such as shown in Fig. 5-17A, the voltage is the same across each component; i.e., $V_A = V_R = V_L$. This value, therefore, is the reference vector in a parallel circuit. I_R is in phase with V_A. I_L is 90° out of phase with the applied voltage and lags the voltage. The sum of the branch

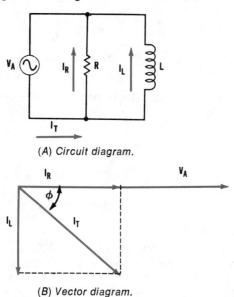

(A) Circuit diagram.

(B) Vector diagram.

Fig. 5-17. Parallel RL circuit.

currents in a parallel dc circuit equals the total current. This is true instantaneously in the ac circuit. Therefore, these effective currents must be added vectorially. The vector I_T represents this vector sum, and ϕ represents the generator phase angle.

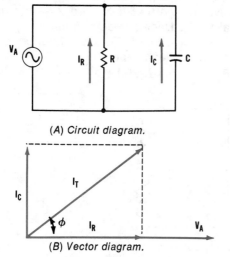

(A) Circuit diagram.

(B) Vector diagram.

Fig. 5-18. Parallel RC circuit.

Parallel RC Circuit

In Fig. 5-18, an RC circuit is analyzed. Since this is a parallel circuit, V_A is again the reference vector. I_R is in phase with V_A. In a capacitor, I leads V by 90°. This is indicated in the vector diagram (Fig. 5-18B). I_T is the vector sum of I_R and I_C. The phase angle is again indicated by ϕ. You will notice that in the parallel circuit, $I_T = \sqrt{I_R^2 + I_X^2}$, whether I_X is capacitive or inductive. Since this is true, you will note that an impedance vector diagram cannot be used since I_T and Z are inversely, not directly, proportional. The simplest way to find the impedance is by the use of Ohm's law; i.e., $Z = V_A/I_T$.

Parallel RLC Circuits

Fig. 5-19A shows the schematic diagram of a parallel RLC circuit. Voltage is the reference vector in a parallel circuit. I_R is in phase with the applied voltage. I_C leads this voltage by 90°, while I_L lags by 90°. Adding vectors I_C and I_L produces a vector that is the difference in their magnitudes and has the direction of the larger. When this vector is added to I_R, I_T is the resulting vector. In the vector diagram shown (Fig. 5-19B), vectors I_C and I_R were added first. The results, however, are the same regardless of which vectors are added first. A lagging phase angle has resulted in Fig. 5-19 since I_L is greater than I_C. If I_C were of a greater magnitude than I_L, the phase angle would lead. If I_C were equal to I_L, then I_T would equal I_R. This condition occurs at the resonant frequency of the circuit, at which current is at its minimum. If current is minimum, then impedance Z is maximum. Notice that a series resonant circuit offers a minimum impedance at the resonant frequency

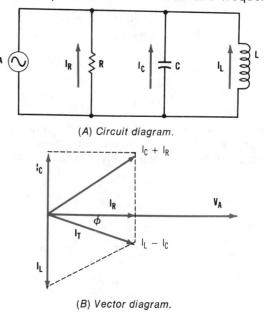

(A) Circuit diagram.

(B) Vector diagram.

Fig. 5-19. Parallel RLC circuit.

while a parallel resonant circuit offers a maximum impedance. At frequencies other than the resonant frequency, an RLC circuit can be either capacitive or inductive. This depends on the relative sizes of L and C.

PHASE RELATIONSHIPS IN THE NULL DETECTOR

The condition for rotation of the balance motor has been stated. It has been stressed that a phase shift of 90° between the current in the line-phase winding and the control winding is necessary for rotation of the balance motor. An understanding of the action of the two-phase motor under such conditions will help us to understand the importance of "phasing" in the null detector. Fig. 5-20 shows the four possible conditions that might exist in the balancing motor. Line voltage is always applied to the line winding of the balance motor. Current, therefore, is always conducting in this winding and its phase is constant, as indicated in the diagram. A thorough analysis of a two-phase motor is more complex than the discussion that will be used here. Only the basic characteristics of the balance motor need to be analyzed for an understanding of the null detector.

Diagram A in Fig. 5-20 shows only the effect of the line current on the motor. This situation would also occur if there were no outputs from the amplifier or if the amplifier output has no 60-Hz component. At the beginning of a cycle of current applied to the line winding of the motor, current is rising. We will arbitrarily assign the polarity of the magnetic field produced by this current as shown in Diagram A. This changing current induces a field into the squirrel-cage rotor of the motor. This rotor field is in the same direction as the field produced by the line winding. Since the north pole of the coil and the south pole of the rotor attract, no torque is produced and no rotation results. As the line current varies, the induced rotor field varies in step with it.

In Diagram B, a voltage is applied to the control winding. This current is in phase with the line current and at the same frequency. As the current in each winding rises, the same magnetic field is produced by each winding. The induced field in the rotor is the result of both of these windings and will be somewhere in between the two. Its actual location will be determined by the relative magnitudes of the two currents. Since the rotor field is equally attracted by the field of both windings, no torque is produced.

Diagram C indicates the situation that exists when the line current leads the control current by 90°. At the beginning of a cycle of current in the line winding, current is rising. The induced rotor field is the same as in situation A. Current in the

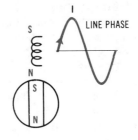

CONTROL PHASE A

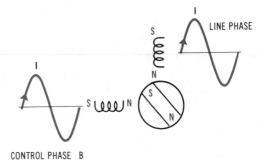

CONTROL PHASE B

CONTROL PHASE C

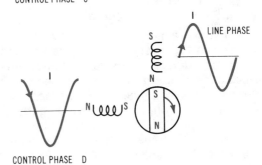

CONTROL PHASE D

Fig. 5-20. Effect of phase on balance motor.

control winding at this instant is at its maximum negative value. No change in this current is occurring at this instant: control current induces no field in the rotor. As the line current continues to rise, however, its rate of change decreases. The current in the control winding begins to rise at this time. Since control current is rising, its magnetic field polarity is the same as that of the line winding. Since its field strength is rising while that of the line winding is decreasing, the rotor field is attracted more and more in the direction of the control winding, resulting in a torque in the counterclockwise direction that starts rotation. As the currents vary, this rotation will continue.

Diagram D indicates the situation that exists when line current lags the control winding current by 90°. Clockwise rotation is the result of these currents. When the line current rises, the control current starts to fall. This means that the polarity of the magnetic field produced by the control winding is opposite to that of the line winding. This results in a repulsion-attraction between the field of the control winding and the rotor field induced by the line current in such a direction as to produce a clockwise torque, resulting in clockwise rotation.

It has been shown that a 90° phase shift between the two phase windings of the balance motor must be produced to create motor control, but how is this phase difference accomplished? In order to answer this question every reactive component in the null detector should be analyzed. We will find that most of these components have very little effect on the phase relationships. We will investigate the converter drive coils, input transformers, coupling capacitors, control windings, and line phase windings to find out which ones affect the phase relationships. In all cases we will use the line voltage for our reference. Since the phase characteristics are a little more involved in the Honeywell ElectroniK 15 recorder, we shall discuss it first, keeping in mind that the same thing may be accomplished in recorders made by other manufacturers.

Phase Relationship of Honeywell ElectroniK 15 Recorder

Fig. 5-21 shows a schematic of the two coil windings of the servomotor of the Honeywell ElectroniK 15 recorder. The control winding has C_{17} of the power amplifier in parallel with the winding. This forms a parallel LC circuit with practically no shift in phase occurring. The line winding, by comparison, has capacitor C_{16} in series with the winding. This series combination causes the phase of the line winding to shift 90°. The shift in phase will be of a constant value and serves as the reference.

The phase of the control winding is subject to a shifting effect depending on the polarity of the dc input signal to the amplifier. If the input is of a positive dc value, the control winding will lag the line winding by 90°. This will cause the motor to rotate in a clockwise direction. When the polarity of the dc input is negative, the control winding will lead the line winding by 90°. This causes the motor to rotate in a counterclockwise direction until balance occurs. Ultimately this means that the control signal shifts phase according to the input polarity, while the phase of the line winding remains at a constant position. As a result

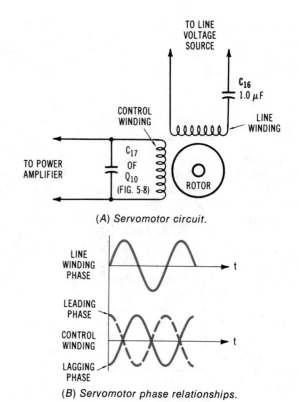

(A) Servomotor circuit.

(B) Servomotor phase relationships.

Fig. 5-21. Servomotor phase relationships.

of this action, the servomotor responds to any polarity change that takes place in the input signal.

Phase Relationships of Mechanical Chopper Recorders

For recorders that employ a mechanical chopper, there are some other phase considerations to take into account. Fig. 5-22A shows a schematic representation of the chopper drive coil, and Fig. 5-22B shows a vector diagram indicating its effect on phase relationships. An applied voltage of 68 volts ac is delivered to the drive-coil circuit by the power transformer. Since this is a series circuit, current will be used as a reference. This is the reference we wish to use since it is the current through the drive coil that controls the vibrating reed. The circuit is composed of R_x and the converter drive coil, which has an inductive reactance, X_L, and a resistance, R. The large resistor (R_x) inserted in the circuit not only drops the ac voltage applied to the circuit from 68 volts to 19 volts, it also reduces the phase angle. In the simplified vector diagram (Fig. 5-22C), V_R and V_{Rx} have been added. They can be added directly since these voltages are in phase. The voltage drops across the resistances are so much larger than that across X_L that the phase angle is very small (10°). An additional lag of 8° is due to contact closure time in the converter. This makes a total phase difference of 18° that can be attributed to the drive coil.

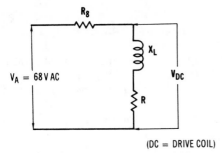

(A) Schematic representation of drive coil.

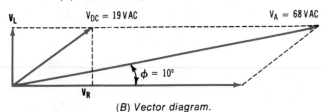

(B) Vector diagram.

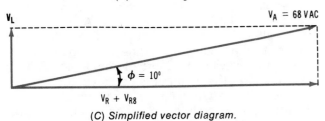

(C) Simplified vector diagram.

Fig. 5-22. Drive-coil phase relationships.

The effect of the input transformer is negligible. This is usually due to the way that such transformers are wound. A transformer can be connected so that its output voltage is either in phase or 180° out of phase with the primary voltage. Either connection could be used here since a 180° phase shift will not produce rotational torque. We have also noted that each transistor produces a 180° phase shift. Again, this will not produce rotational torque in a two-phase motor.

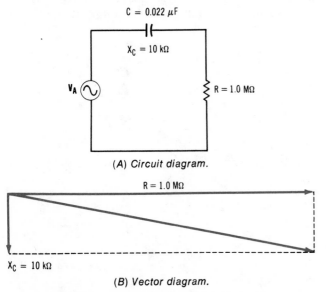

(A) Circuit diagram.

(B) Vector diagram.

Fig. 5-23. Phase shift in a coupling network.

The next reactive component to be considered is the coupling capacitor. These components also have very little effect on amplifier phase relationships. Fig. 5-23A shows a typical RC coupling network. With a typical coupling capacitor (C = 0.022 μF) used in the amplifier at a frequency of 60 Hz, X_C will be,

$$X_C = \frac{1}{2\pi fC}$$

$$= \frac{1}{6.28 \times 60 \times 0.022 \times 10^{-6}}$$

$$= \frac{0.159}{1.32} \times 10^6$$

$$= 120\,000\ \Omega$$

This value is about 1/8 the size of R. Vector R, therefore, will be about 8 times as long as X_C in the vector diagram (Fig. 5-23B), producing a phase angle of about 7°. Even when four coupling circuits are used, the total phase shift would be small. Some of the capacitors used for coupling have even less reactance than the one involved here.

The control winding of the balance motor is considered next. It is the output load of the whole amplifier, and the power amplifier in particular. A capacitor is connected in parallel with this winding. These two components form a resonant circuit at a frequency of 60 Hz. This parallel circuit then acts as a resistive load. No phase shift occurs in the control winding.

Most of the phase shift from the line voltage occurs in the line phase winding. This winding is connected to the line voltage, V_A, through a capacitor. The reactance of this capacitor at 60 Hz is much greater than the reactance of the coil winding. The result is that the overall effect of the circuit is capacitive. As seen in Fig. 5-24, 120 volts (V_M) is applied to the phase winding itself. This voltage is the vector sum of the voltage drops across the inductive reactance of the coil and the resistance of the coil. V_M leads I by about 37.5°. This voltage drop (V_M) added vectorially to the voltage drop across the capacitor (V_C) equals the applied voltage, V_A, of 120 volts. This voltage lags the current by 37.5°. The effect of this circuit is that V_A lags V_M by 75°.

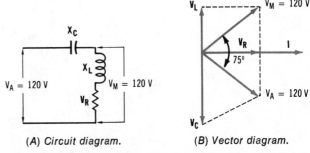

(A) Circuit diagram.

(B) Vector diagram.

Fig. 5-24. Phase relationships in a motor drive circuit.

The combined effect of the phase shift caused by the drive coil (18°) and that of the line winding (75°) is a total of 93°—about the phase shift desired. This difference in phase in the line winding will be constant. The power amplifier output, however, changes its phase 180° with a change of polarity of the dc input voltage. The result is that the amplifier output may lead the line phase by 90° or lag by 90°. This produces the type of motor action necessary for rebalance in the measuring circuits.

SUMMARY

The Honeywell ElectroniK 15 recorder contains two major divisions that are discussed in this chapter. The amplifier has a power supply, filter, chopper, voltage amplifier, and driver stage. The motor unit contains a power amplifier and the balancing motor.

The power supply provides ac to the line winding of the motor and it develops dc for component operation. A full-wave rectifier with a center-tapped transformer and an RC filter provides dc operating voltages for component operation. Three different dc voltages are provided. Two are regulated by zener diodes. In addition to this, a negative dc source is also developed for the wave-shaping circuit.

The input filter is a three-section RC network that filters noise and extraneous ac from the dc input signal.

The chopper changes dc input into ac. This is achieved by a waveshaping transistor that changes ac from the transformer into a pulsating dc. An IGFET has a square wave applied to its gate, the dc slide-wire feedback signal to its source, and the dc input to its drain. The output corresponds to the two signal inputs and alternates between high-resistance and low resistance conduction due to the gate signal.

Voltage amplification is achieved first by a JFET and two bipolar transistors in the input section of the amplifier. Following this, three direct-coupled amplifiers are employed. The collector of one transistor and the base of the next transistor are commonly connected together in this circuit configuration.

The driver/power amplifier is responsible for the output of the amplifier unit. An amplified ac signal processed by the driver is used to control a large amount of current controlled by the power amplifier. The term *null detector* is sometimes used to describe this section of the amplifier.

The servomotor of a recorder is responsible for moving the pen and altering the slide-wire balancing mechanism. The servomotor is basically a two-phase induction motor. The ac applied to the line winding has a fixed phase relationship, while that applied to the control winding shifts according to the polarity of the dc input. A positive dc input will produce a lagging phase of 90° with respect to the phase of the line winding. This will cause a clockwise rotation of the motor. When the polarity of the dc input is negative, the control winding will lead the line winding by 90°. This, in turn, causes counterclockwise rotation of the motor.

When ac is applied to a resistor, the current and voltage remain in phase. In an inductive circuit with ac applied, the current lags the voltage by 90°. A capacitive circuit causes the current to lead the voltage by 90°. When R, C, or L components are connected in series, the phase relationship between I and V is dependent upon the predominant component value. In a circuit containing parallel components, the voltage serves as the reference component with the current either lagging or leading, depending on the predominant component. The resulting effect of current is the vector sum of the individual component current values.

CHAPTER 6

Taylor Transmitter

INTRODUCTION

In many electronic process-control applications, the signal may go directly to the system controller, indicator, or recorder without any further signal processing. In other applications, controllers may be positioned at some remote location from the process variable being manipulated. In this case, special pieces of interface equipment are needed to receive the process signal from the sensor and give it the power needed for transmission to the controller. In achieving this function, the signal must be strong enough to travel the required distance without excessive noise or any reduction in accuracy or sensitivity of the signal. Signal transmitters are used to achieve this function in electronic instruments.

A simplified block diagram of the Taylor Model 1021T millivolt-to-current transmitter is shown in Fig. 6-1. This instrument has a number of unique things in its circuitry that have not been discussed previously. For example, it has electrical feedback rather than the electromechanical devices used in some of the preceding units. Input and output isolation is accomplished optically. Isolation between electrical power and the input/output circuitry is also provided on all units as standard equipment. When used in thermocouple input applications, a cold junction is self-contained in a constant-temperature oven. Transmitters of this type have high reliability and are immune to radio frequency (rf) interference and ambient temperature variations.

POWER SUPPLY

The power supply of Fig. 6-2 shows one of the unique options of the transmitter. In this case, 24 volts dc is used as the primary input source. In order to isolate the unit from the primary electrical sources and to produce higher voltages, a dc-to-ac converter is used in the primary circuit. This circuit causes the primary dc voltage to be switched on and off through the two transistors. As a result of this action, the magnetic field of the transformer expands and collapses with each switching change.

When dc power is applied to the two transistors, more current will conduct through one transistor than the other due to a slight difference in the transistors and the tolerance values of the components. Assume in this case that Q_{100} is initially conducting more heavily than Q_{101}. This will result in a base-drive voltage being induced into the lower winding of transformer T_{100} of such a polarity as to cause the collector current of Q_{100} to increase and the collector current of Q_{101} to decrease. As a result, Q_{100} is driven into saturation and Q_{101} is driven to cutoff. At the saturation point of Q_{100}, there is no further change in collector current and no further voltage induced into the lower winding of transformer T_{100}. This means that the cutoff voltage to the base of Q_{101} is removed, which allows it to begin conduction. Through transformer action, this now causes an increase in the collector current of Q_{101} and a decrease in the collector current of Q_{100}. Ultimately, Q_{101} saturates and Q_{100} goes into cutoff. The process then repeats itself with Q_{100} conducting again and Q_{101} being cut off again.

The switching action of Q_{100} and Q_{101} produces a varying voltage across the primary winding of transformer T_{100} which is either stepped up or down by the secondary windings. The secondary windings of this circuit are connected to typical bridge circuits with RC filtering. Through the action of this dc-to-ac converter power supply, it becomes possible for this transmitter to work from a common dc source and still have electrical isolation.

VOLTAGE REGULATION

In conjunction with the power supply, the Taylor Model 1021T has a number of regulated output voltages. Fig. 6-3 shows a simplified schematic

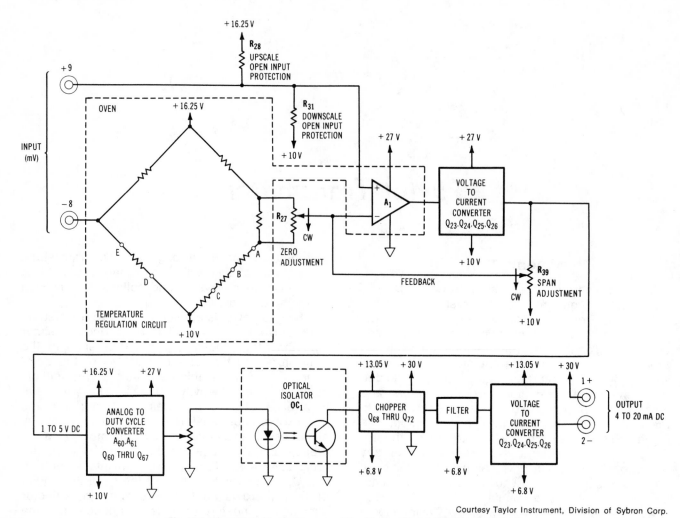

Courtesy Taylor Instrument, Division of Sybron Corp.

Fig. 6-1. Simplified block diagram of the Taylor Model 1021T transmitter.

of the voltage regulator for the +30-volt, +13.05-volt, and +6.8-volt dc supplies. Through this type of circuitry, the dc voltage will change very little regardless of variations in load or line voltage. Regulation in this case is achieved by zener diodes.

A zener diode, as you will note, is connected in reverse bias. The cathode, which is representative of the n material, is connected to the positive side of the dc source, while the anode, or p material, is connected to the negative side. When the reverse-bias voltage exceeds the zener voltage (V_Z), the diode goes into conduction. When this occurs, the voltage across the diode remains at its rated V_Z value. A series-connected resistor absorbs the remaining voltage through an increase in current.

In Fig. 6-3, it can be seen that the positive output of the power supply is connected to the cathodes of D_{64} and D_{66} through a 1000-ohm resistor and a 499-ohm resistor. When the power supply is turned on to produce voltage, 30 volts appears at the output immediately. With the zener

voltage of D_{64} at 6.25 volts and D_{66} at 6.8 volts, these voltages will appear across the respective diodes. Note also that 6.25 volts plus 6.8 volts equals 13.05 volts, which appears across the two diodes. The difference between +30 volts and +13.05 volts is +16.95 volts, which is the combined voltage that appears across R_{79} and R_{80}.

Any change in load or line voltage that reduces the cathode voltage of D_{64} or D_{66} reduces the total conduction current of the series circuit. This reduction in current decreases the voltage drop across R_{79} and R_{80}, which compensates for the reduction in voltage. Any rise in voltage across the circuit increases the diode current. This results in an increased voltage drop across R_{79} and R_{80}, which maintains the voltage across the diodes at the rated V_Z value.

The zener voltage of the two diodes maintains a constant value within an operating range of current depending on the wattage rating of the diodes. A 1-watt 6.8-volt zener diode, for example, would accommodate up to $I = P/V = 0.1471$ amperes, or 147.1 milliamperes of current. When

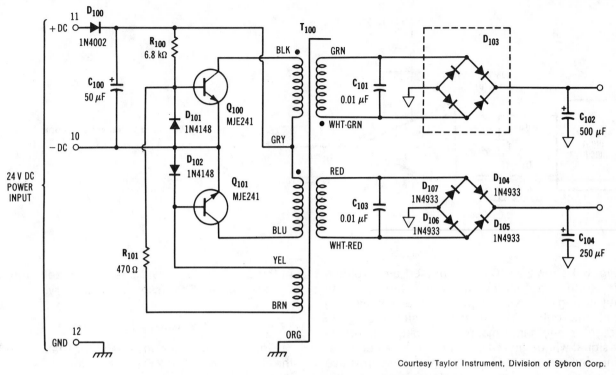

Courtesy Taylor Instrument, Division of Sybron Corp.

Fig. 6-2. Power supply for the Taylor transmitter.

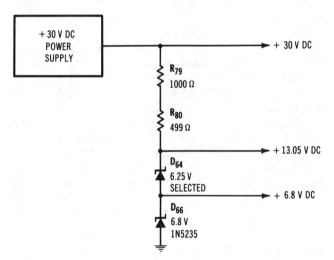

Fig. 6-3. Simplified schematic of a voltage-regulator circuit.

diodes are connected in series, as they are in Fig. 6-3, both diodes must have a similar current-handling capability or wattage rating.

CONVERTER CIRCUITS

The block diagram of the Taylor Model 1021T transmitter of Fig. 6-1 should have indicated an additional feature of the unit with respect to converter circuitry. Previous units, for example, employed some type of mechanical dc-to-ac converter in the signal path to produce ac for easy signal processing. A converter drive coil, mechanism,

and transformer were needed to accomplish this operation. The Taylor transmitter is unique in this regard because it does not employ a mechanical dc-to-ac converter in its signal path. It simply accepts a low-voltage dc signal in the millivolt range and processes it through direct-coupled amplifiers to produce a suitable dc output current. In this unit, dc is processed by not employing reactive components such as transformers or capacitors in the signal path. With the aid of operational-amplifier ICs, dc amplification is much easier to achieve than it was in older vacuum-tube transmitters.

An electronic form of signal conversion is, however, used in the Model 1021T transmitter for a different purpose. The circuit in this case is called an analog-to-duty-cycle converter. Its function is to change the variable value of dc voltage applied to its input into a square wave. The on, or duty-cycle time, of the square wave varies according to the value of the applied dc voltage. The output of the converter is then applied to the input of the optical isolator circuit.

VOLTAGE AMPLIFIERS

A great deal of the voltage amplification achieved by the Taylor transmitter is accomplished by dc operational-amplifier ICs. In this circuit application, a dc voltage value is applied to the input and an amplified version of dc appears at the output.

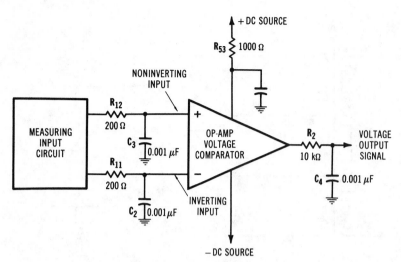

Fig. 6-4. An op-amp voltage comparator.

Fig. 6-4 shows an IC op amp used as a voltage comparator. In this circuit, when the inverting input is negative with respect to the noninverting input, the output produces a positive-going dc voltage. In the same manner, a positive-going input signal will be inverted and produce a negative-going output signal. When negative or positive signals are applied to the noninverting input, the output will be an amplified version of the input with no change in polarity.

The input of the op amp shown is connected to the output of a Wheatstone bridge circuit. In this application, the amplifier responds to very small changes in dc voltage produced by the input signal. A minute change in input signal voltage will produce a high level of voltage amplification through the IC. Gain capabilities of several thousand are not unusual with this type of amplifier circuit. Op amps used in this type of circuit do not employ an external feedback resistor between the input and output that restricts amplification. Since there is no intervening reactive component

in the signal path, dc will pass very readily through the op amp and its associated circuitry.

OPTICAL ISOLATORS

Fig. 6-5 shows an optical isolator that is used in the Taylor Model 1021T transmitter. The isolator is a unique circuit innovation that provides isolation between the input signal and the output or load device. Essentially, the amplified dc input signal is applied to a circuit that changes dc voltage of a variable value into a square wave. This analog-to-duty-cycle converter alters the on and off time of the square wave in accordance with the voltage value of the dc input signal. The square wave is then applied to a light-emitting diode (LED) in the optical coupler. Variations in light intensity produced by the LED are then picked up by the base of a phototransistor housed in the same enclosure. The light-signal path from LED to phototransistor can only occur in one direction, which assures isolation between the input and output.

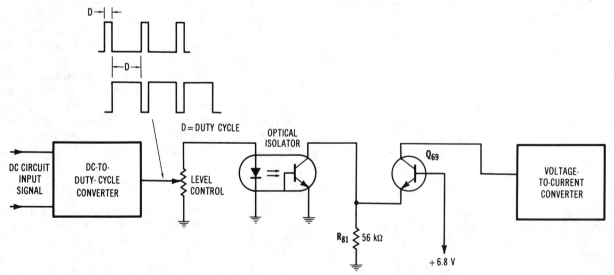

Fig. 6-5. Simplified schematic of an optical-isolator circuit.

The output of the phototransistor of the optical isolator must now be changed from a square wave into a variable dc voltage before it is applied to the output circuit. Initially, the square-wave signal is applied to a chopper circuit. The square wave simply turns a transistor on and off according to the duty cycle of the square wave. The output of the chopper is then filtered and applied to a voltage/current converter circuit. Ultimately the signal is applied to the current output amplifier circuit for further processing.

DARLINGTON AMPLIFIER

Fig. 6-6 shows a schematic of two transistors connected into a Darlington-pair circuit configuration. A circuit of this type develops the output signal for the Taylor transmitter. A Darlington amplifier has a high input impedance, a low output impedance, and very high current gain. The voltage gain of the circuit, however, is less than one. All of these features are desirable characteristics for the current amplifier of the Model 1021T transmitter.

A Darlington pair is often referred to as a double emitter-follower circuit configuration. The input in this case is applied to the base of Q_1, and the output is taken from the emitter of Q_2. The combined transistor pair has slightly less than unity voltage gain. The total current gain, however, is the effective product of the current gains (beta, or

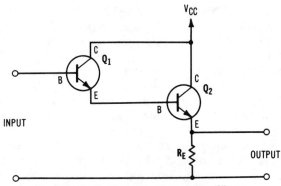

Fig. 6-6. Darlington-pair current amplifier.

h_{fe}) of the two transistors. The input impedance is increased because of this current multiplication and is essentially the total current gain times the value of the emitter resistor, R_E.

A wide variety of Darlington amplifiers are built on a single chip and housed in the same case. This type of configuration behaves in the same manner as a single transistor emitter-follower amplifier with a very high current gain. Devices of this type are designed to simplify circuit construction techniques.

MEASURING CIRCUITS

Fig. 6-7 is a simplified schematic of the measuring circuit of the Taylor transmitter. We have in this circuit the Wheatstone bridge, zero adjust-

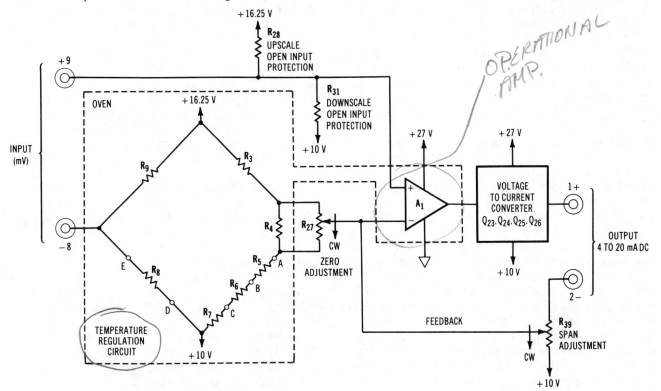

Courtesy Taylor Instrument, Division of Sybron Corp.

Fig. 6-7. Simplified block diagram of the Taylor Model 1021T measuring circuit.

ment, downscale or upscale open input protection circuits, temperature regulation, cold junction block amplifier, and the span-adjustment control. The input to the unit is a low-level signal of a few millivolts. The combination of jumper connections between terminals A through E on the bridge are used to position the range of the zero adjustment determined by the value of the input signal from −25 to +50 millivolts dc.

The broken line surrounding the bridge circuit components indicates those things that are temperature sensitive and located in the oven. The temperature of the oven is controlled by a temperature regulating circuit consisting of a thermistor sensor and a transistor. Heat from the transistor is used as a thermal source for the enclosed oven. The thermistor is a sensitive resistor that changes its resistance value with temperature. Through this resistance, the base current of the transistor is varied accordingly.

The voltage source for the bridge circuit is +16.25 volts and +10 volts from the regulated source. The +16.25-volt source is connected to the junction of resistors R_3 and R_9, while the +10-volt source is applied to the junction of R_7 and R_8. Resistor R_{27} is a zero-adjust potentiometer located outside of the oven assembly. When the bridge is balanced to zero by R_{27}, there should be approximately 0.5 milliampere of current in each leg of the bridge. When the bridge is balanced, the voltage drop on one side should equal the voltage drop of the alternate side. When this occurs, there should be zero voltage input to the inverting input (−) of the op amp. The cold block junction is attached to the positive input and applied to the noninverting input (+) of the op amp. Variations in temperature are compensated for through this connection.

When the signal applied to the input of the bridge circuit increases, it alters the input voltage of op amp A_1. An increase in input voltage causes a corresponding increase in output voltage. The output of A_1 is then used to control the voltage signal applied to the voltage-to-current converter. The output of the converter provides from 4 to 20 milliamperes of dc for the instrument load.

A portion of the output signal forms a feedback loop that returns to the inverting input of amplifier A_1. When the value of the input signal equals the output feedback signal applied to A_1, the transmitter becomes stabilized at a new value.

The transmitter also has provision for upscale and downscale open-input protection. Upscale open-input protection is achieved by connecting an 82-megohm resistor (R_{28}) to the +16.25-volt dc source and the noninverting input of amplifier A_1. In normal operation, the input signal and the bridge circuit are shunted across resistor R_{28}, which makes it inoperative. If by chance the input circuit becomes open, R_{28} provides a voltage signal to the input of amplifier A_1 which causes the output of the transmitter to drift upscale.

Downscale open-input protection is provided by a 22-megohm resistor (R_{31}). This resistor is connected to the +10-volt dc source and the noninverting input of A_1. In normal operation, the input signal and the bridge circuit are shunted across resistor R_{31}, which makes it inoperative. If by chance the input circuit becomes open, R_{31} provides a voltage signal to the input of amplifier A_1 which causes the transmitter output to drift downscale.

SUMMARY

Signal transmitters are designed to receive the process signal from the sensor and give it the power needed for transmission to a controller located some distance away. Typically, a transmitter employs temperature regulation, voltage-to-current conversion, an analog-to-duty-cycle converter, optical isolators, a chopper, filtering, and a final voltage-to-current converter.

The power supply of the Taylor unit employs 24 volts dc as a primary power source. A dc-to-ac converter is used in the power supply of the transmitter discussed in this chapter. With two transistors, one conducts while the other is cut off. The process then reverses, which causes dc to be changed into a switching on and off action. Applied to a transformer, this voltage can be stepped up or down accordingly. In a sense, the power supply changes dc to ac, steps it up, rectifies the ac to dc, and filters the dc.

Zener diodes are used to regulate dc power-supply voltage values. These devices are connected in a reverse-bias direction. Changes in the input voltage or load resistance values will not alter the regulated output voltage within the operating range of the regulator.

One form of conversion used in the Taylor transmitter is called analog-to-duty-cycle conversion. This function changes variable dc values into square waves. The on part of the square wave, or its duty cycle, is made to change according to the value of the applied dc.

The voltage amplification of the Taylor transmitter discussed in this chapter is achieved by operational amplifiers. Gain capabilities of several thousand are typical with IC op amps. With these devices, dc can be amplified because no intervening reactive components are involved. The op-amp circuits discussed here have high amplification capabilities because no feedback resistor is used to restrict gain.

Optical isolators are used to provide isolation between the input signal and the output or load device. First, ac is changed into a square wave with a variable duty cycle. This signal is then applied to a light-emitting diode in the optical coupler. Changes in light intensity are then picked up by the base of a phototransistor housed in the same enclosure. The signal path can only occur between the LED and the phototransistor, which assures isolation between input and output. The output of the phototransistor is then changed into a variable dc value before being applied to the output circuit.

A Darlington amplifier is two transistors con-nected in a double emitter-follower configuration. It has high input impedance, low output imped-ance, and high current gain.

The measuring circuit employs a Wheatstone bridge, zero adjustment, downscale or upscale open-input protection, temperature regulation, cold block junction amplification, and the span-adjustment control. The input is normally only a few millivolts. The voltage source for the measur-ing circuit is supplied by a regulated section of the power supply.

pH Analyzers

INTRODUCTION

It is extremely important in many chemical process applications to know if a particular chemical solution has a predominantly acid or alkaline (base) content. Some common examples of acid solutions are vinegar (acetic acid), the citric acid in fruit juice, and dilute sulphuric acid, which is used as a battery electrolyte. Ammonia water, by comparison, is a rather weak base solution, while concentrated lye mixtures form a very strong base solution. Acid and base have entirely different chemical reactions when they exist in solutions. Successful control of chemical processing, therefore, necessitates that acid and base levels be carefully controlled to assure a desired outcome.

A more specific definition of pH refers to the number of ionized or free hydrogen ions (H^+) and hydroxyl ions (OH^-) in a solution. Acid has an abundance of H^+ ions, while base has large numbers of free OH^- ions. The pH value is, therefore, a measurement of the ratio of hydrogen and hydroxyl ions in a solution. When H^+ is predominant the solution is acid. When OH^- is predominant, the solution is base. If equal amounts of base and acid are present, the solution is a neutral salt.

The numbering system for a pH scale ranges from 0 to 14. The number 7 is at the center of this span and is considered to be an indication of a neutral solution. Acid levels occupy the position from 7 down to 0, with the smaller numbers indicating the highest acid levels. The numbers 7 to 14 represent the base scale, with the largest numbers indicating the highest base levels (see Fig. 7-1).

pH MEASUREMENT

The pH level of a solution can be determined by direct measurement of the dc voltage developed between two electrodes immersed in the solution under test. A dc null-balancing potentiom-

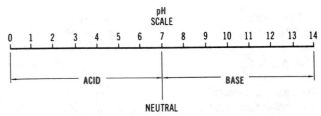

Fig. 7-1. The standard numbering system for a pH scale.

eter with a very high input impedance is commonly used to indicate the voltage developed by the measuring electrode probe. The electronics part of the instrument is then responsible for manipulating the voltage developed by the electrodes in such a way that it will develop an indication of pH. Hand-deflecting instruments, chart recorders, and digital readout displays are in common use today. A number of different display techniques are used in pH measurement. In this chapter (and the next), we will assume that the display device is a hand-deflection meter or a recorder. Recorders obviously provide a permanent record of pH levels for a variety of different time spans.

Fig. 7-2 shows a typical pH indicator scale for a hand-deflection instrument. As you will note, there are ten small graduations or divisions between each two numbers. Each division, therefore, represents 0.1 pH units. If the indicating hand is deflected to the third small graduation to the right of the number 5, the pH level would be 5.3. This indicates an acid level of approximately 20%. A pH meter can also be used to indicate positive and negative voltage values. In this particular indicator, full-scale deflection is +700 to

Fig. 7-2. A typical pH indicator scale.

−700 millivolts. In practice, the deflecting hand should come to rest at zero when measurements are not being taken.

The curved black strip of the pH meter scale of Fig. 7-2 represents a mirror finish. This part of the meter scale is designed to produce a reflection of the indicating hand. In practice, an operator looks at the scale in such a way that the hand and its reflection are exactly in line (i.e., the reflection cannot be seen). When this occurs, the indication should be quite accurate because the operator is not looking at the scale from an angle. Technically this is described as reducing meter parallax.

pH PROBES

The probe or electrode part of a pH instrument is often thought of as a battery whose voltage varies with the pH level of the solution in which it is placed. It contains either two separate electrodes in a single probe housing or two distinct probes. In either case, one electrode, or part of the common probe, is sensitive to hydrogen. A special glass bulb or membrane material is used that has the ability to pass H^+ ions inside of the sensitive bulb. When the electrode is placed in the solution, a voltage that is proportional to the hydrogen ion concentration is developed between an inner electrode and the outer electrode or glass bulb material. This pH-sensitive electrode is often called a *half-cell*.

A second discrete electrode, or the alternate part of the common probe, is used to develop a reference voltage. The reference electrode is primarily responsible for producing a stable voltage that is independent of solution properties. This electrode, or half-cell, develops a fixed voltage value when placed in the solution. When the reference half-cell and the pH glass-bulb half-cell are combined, they form a complete probe. Fig. 7-3 shows a typical pH probe.

pH INSTRUMENTATION

Fig. 7-4 shows a general diagram of the essential parts of a pH measuring instrument. In this case, pH is observed on a hand-deflection meter scale. Quantitative measurements of pH are produced by this type of instrument. A majority of the pH instruments used in industry today are this type. They range from small portable units housed in convenient carrying cases to larger stationary units.

The essential parts of a pH instrument, regardless of its type or style, are the measuring half-cell electrode, the reference half-cell electrode, a high-impedance amplifier, and an indicator. This type of instrument is normally classified as a di-

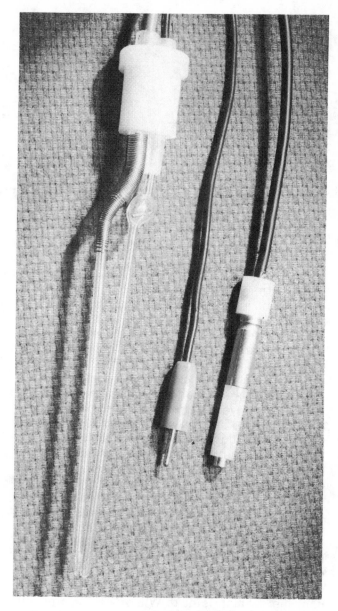

Courtesy Leeds & Northrup Co.

Fig. 7-3. A typical pH probe.

rect-reading unit because it responds to voltage values produced directly from the solution under test. Nearly all industrial pH instruments are of this type as opposed to the indirect method that produces an indication due to color changes in a material sample.

The measurement of pH presents some interesting electronic circuits that have not been previously discussed. Fig. 7-5 is a block diagram of a typical pH indicator that shows some of these items. One of the new problems is the very high impedance of the measuring probe, which makes many different design features necessary. A negative feedback measuring circuit, which is similar to a null-balancing system, is one of these necessary features. Careful shielding of the input cir-

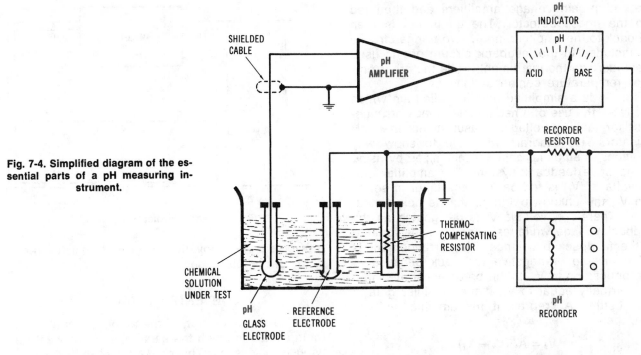

Fig. 7-4. Simplified diagram of the essential parts of a pH measuring instrument.

cuits is also necessary. An emitter-follower amplifier is likewise needed to match the high impedance of the input circuit to the low input impedance of a high-gain solid-state voltage amplifier.

A great deal of care must also be taken to keep the amplifier circuit from oscillating. Any ac output of the amplifier must be fed back so as to be 180° out of phase with the ac present in the amplifier. Of course, the simplest way to suppress oscillation is to keep the ac voltage in the feedback loop to a minimum. Some of the methods used to do this have already been mentioned. Shielding, for instance, helps keep spurious ac voltages out of the amplifier. Using all dc source voltages helps to reduce this problem. All ac voltages, unfortunately, cannot be kept out of the

feedback loop. Oscillation can be prevented, however, if the ac signal is fed back out of phase, but circuits must be provided for this purpose.

The feedback measuring circuit has another purpose besides keeping the input impedance high. It also helps to counteract fluctuations in amplifier gain due to circuit variations, line-voltage changes, etc., as long as the amplifier gain is high. From the block diagram, it can be seen that the electrode voltage is fed to the converter circuits. This dc voltage is changed to an ac voltage by a vibrating-reed type of converter. The ac signal is then fed to an emitter follower, which is the first stage of the voltage amplifier. As mentioned before, this circuit is for impedance-matching purposes. The output is amplified by a number of discrete tran-

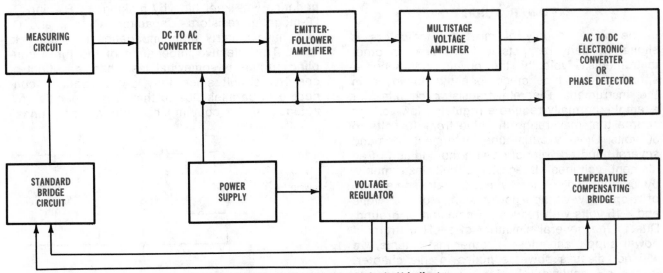

Fig. 7-5. Block diagram of a typical pH indicator.

sistor high-gain voltage amplifiers and then fed into the phase detector. The dc output is then fed back to the amplifier input. Two bridge circuits are included in the feedback circuit; one is used to standardize the instrument, while the other provides temperature compensation.

Fig. 7-6 is a simplified amplifier diagram which illustrates the use of a negative feedback arrangement for making voltage measurements in a pH indicator. This diagram will help to show why variations in amplifier gain are minimized by using the negative feedback system. The amplifier output voltage (V_f) is fed back to the input in series with V_x, the unknown voltage that is being measured. Therefore, voltage V_f is designated as the feedback voltage, while resistance R_f is designated as the feedback resistance. This arrangement is called inverse or negative feedback because V_f is in opposition to V_x. It can be demonstrated that V_f is virtually equal to V_x, if the amplifier gain is high. Letting A_V represent the amplifier voltage gain, then

$$V_f = A_V (V_x - V_f)$$

from which we obtain

$$V_x = V_f \left[1 + \left(\frac{1}{A_V} \right) \right]$$

so that if A_V is sufficiently high, then $1/A_V$ is very small and can be neglected. It then follows that

$$V_x \cong V_f$$

This indicates that the operation of the circuit is not dependent upon the exact value of the amplifier gain, but only requires the gain to be high. Therefore, gain fluctuations caused by line voltage changes, circuit variations, etc., have very little effect on measurement indications as long as the gain remains high.

POWER SUPPLY

The power supply of most pH indicators is similar in many respects to those used in other instruments. Typically, the power supply corresponds to the type of components employed in the instrument. Discrete transistor circuits, for example, normally require a regulated low-voltage source that may range in value from 10 volts to 50 volts. Older vacuum-tube instruments demand several hundred volts of dc with the addition of ac filament voltages. Integrated-circuit instruments, by comparison, generally necessitate some type of a split low-voltage power supply of +15 volts and −15 volts with respect to common or ground. Due to the general similarity of a pH instrument power supply and those of other instruments, we will not discuss power supplies in this chapter. In the pH analyzer chapter, however, a unique

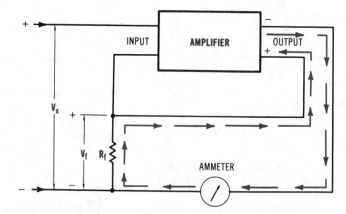

Fig. 7-6. Feedback measuring circuit.

power supply innovation using a transistor regulator will be discussed.

CONVERTER CIRCUITS

Dc-to-ac conversion of the amplifier input is accomplished in much the same manner as was previously discussed. The converter circuit is shown in Fig. 7-7. Two low-impedance features of the previous converter circuits had to be eliminated, however. The first change involved the chopper itself. In previous amplifiers, the chopper has been normally closed. By this we mean that the contacts were closed more than 50% of the time. The pH indicator chopper, on the other hand, is normally open. The switch contacts are actually closed only a very short portion of the operating cycle. This means that current occurs only for a very short percentage of each cycle. Average current, therefore, is very small. As a result, the converter acts as a high impedance.

The low-impedance effects of an input transformer also have to be eliminated. In its place, three coupling networks were installed. The impedance was kept high by making R_2, R_3, and R_4 20-megohm resistors. Capacitors C_2, C_3, and C_4 also have a fairly high impedance to the 60-Hz input. The effective impedance of the three coupling circuits is somewhat less than the effect of one RC circuit since they are essentially connected in parallel. Due to the large capacitive reactance of the coupling capacitors, some phase

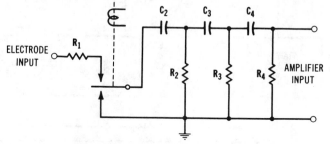

Fig. 7-7. Dc-to-ac converter circuit.

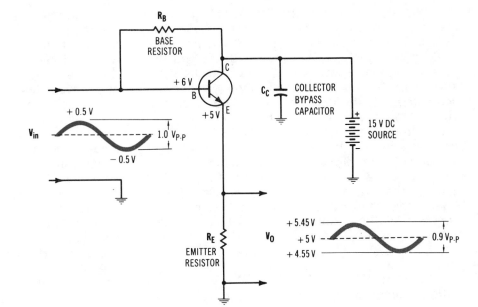

Fig. 7-8. Emitter-follower amplifier circuit.

shift occurs in the amplifier input. This will be discussed in the section on phasing. Resistor R_1 is 1000 megohms and also adds to the input impedance.

EMITTER FOLLOWER

A schematic diagram of a typical emitter-follower amplifier is shown in Fig. 7-8. The collector of this circuit, as you will note, does not have a load resistor. If a collector resistor is used, however, a bypass capacitor returns the signal to ground. This capacitor effectively grounds the collector ac voltage. In some references, emitter-follower amplifiers are called common-collector or grounded-collector amplifiers because of this ground return connection.

Since there is essentially no collector load resistor in an emitter-follower amplifier, the only output load is the emitter resistor, R_E. The resistance value of R_E is largely responsible for the output impedance of the amplifier. In practice, R_E is kept low which means the output impedance is likewise low. With the input impedance of a typical amplifier high, a unique characteristic of the emitter-follower amplifier is high input impedance and low output impedance. Since the output voltage developed across the low-impedance emitter resistor is small, emitter followers are known to have a voltage gain of less than one. The power gain and current gain of this amplifier, however, may range as high as 100 in a representative circuit. Impedance matching and power amplification are the primary applications of an emitter-follower amplifier.

The emitter-follower amplifier normally has only a few volts developed across the emitter resistor with no signal applied. This voltage, as indicated in Fig. 7-8, is positive with respect to ground. When the base signal goes positive, it causes a corresponding increase in emitter current through R_E. This, in turn, means an increase in positive emitter voltage. In a similar manner, when the base signal voltage swings negative, it causes a decrease in emitter current. This means a corresponding decrease in the emitter voltage. In a sense, the emitter voltage follows the value change in base input voltage. This is where the circuit derived the name emitter follower. This relationship between input and output voltage can also be described as being in-phase.

For purposes of illustration, let us again look at the emitter-follower amplifier of Fig. 7-8 with no signal applied. In this case, note that the emitter voltage is +5 volts and the base voltage is +6 volts. The base is, therefore, 1 volt positive with respect to the emitter. This adequately assures a suitable level of forward biasing. Assume now that the amplifier also has a voltage gain factor of 0.9.

When the positive 0.5-volt half-cycle of the ac input signal is applied to the base, it will cause a corresponding increase in the emitter-collector current. With a gain of 0.9, a +0.5-volt input signal will cause the emitter voltage to increase by 0.45 volt, or 5.0 volts + 0.45 volt = 5.45 volts. With the original no-signal base-emitter voltage of 1 volt, this change in signal causes the emitter-base voltage to now become 6.5 volts − 5.45 volts, or 1.05 volts. This means that a 0.5-volt input only causes a 0.05-volt change in the emitter-base voltage.

In the same manner, when the negative 0.5-volt half-cycle of the ac signal is applied to the base-emitter junction, it will cause a corresponding reduction in the emitter-collector current. The voltage gain will be −0.5 volt × 0.9, or −0.45 volt, which will produce an emitter voltage of 5.0 volts

— 0.45 volt, or +4.55 volts. The resulting base-emitter voltage will now be +5.5 volts − 4.55 volts, or 0.95 volt. This means that a 1.0-volt peak-to-peak input signal only causes a 0.9-volt peak-to-peak change (5.45 volts to 4.55 volts) in the voltage across R_E. This effect is commonly called *degeneration.* Emitter-follower amplifiers have an inherent degeneration problem.

Due to the effect of degeneration in an emitter-follower amplifier, the base only draws a nominal amount of current. In transistor circuits of this type, degeneration has a rather significant influence on the total amount of signal gain achieved. Due to this characteristic, emitter-follower amplifiers are not used in applications that demand voltage gain.

VOLTAGE AMPLIFIERS

In pH instruments, voltage signals produced by the measuring circuit generally necessitate a rather high level of amplification in order to produce an output that will drive an indicator. In many instruments, this function is achieved by a combination of two or more discrete component RC-coupled transistor amplifiers. In this type of amplifier, the input signal is applied to the base of the first amplifier and removed from the collector. This signal, which is ac because of the converter, passes through a coupling capacitor to the base of the next transistor. The process continues, base to collector, through succeeding amplifiers.

Fig. 7-9 shows a typical two-stage RC-coupled transistor amplifier. In this circuit, each transistor uses emitter biasing. Transistor Q_1, for example, develops voltage across R_4 when the transistor is conducting. This voltage makes the emitter slightly negative with respect to ground. The base of the transistor is also made negative by a much larger value due to the voltage divider network composed of R_1 and R_2. Through this, the base is more

negative than the emitter, which results in forward biasing. Voltage variations across R_4 due to ac signal changes controlled by Q_1 are bypassed around R_4 by capacitor C_3. This reduces degeneration and maintains the emitter voltage at a constant level. Resistors R_5, R_6, and R_8, and capacitor C_4 achieve the same result for transistor Q_2.

Resistor R_3 connected to Q_1 is the collector load resistor. In practice, the resistance value of R_3 is quite large in order to achieve high levels of amplification. Capacitor C_2 is used to couple the ac output signal of Q_1 to the input of Q_2. Functionally, C_2 must pass the ac signal easily and block the passage of dc. Typical capacitor values are 50 microfarads. C_8 and R_7 of transistor Q_2 achieve the same function. Capacitors C_6 and C_7 are 0.002-microfarad rf noise filters. These capacitors provide a low capacitive reactance to high-frequency ac, and eliminate it by bypassing it to ground. The circuitry of Q_2 is identical to that of Q_1. After the signal has been satisfactorily amplified by the two transistors, the collector load resistor usually decreases in value and the emitter resistor increases. As a general rule, this type of design prevents distortion of the signal due to excessive amplification.

With the availability of high-gain operational amplifiers built on a single IC chip, multistage amplifiers are often replaced today with a single op amp. The Beckman pH analyzers of Chapter 8 will show an application of this type of circuitry. Older vacuum-tube pH indicators use a similar circuit that contains three or more RC-coupled amplifiers. The primary principle of operation is very similar to that of the transistor circuit.

AC TO DC ELECTRONIC CONVERTER

In our representative pH instrument, we started with a dc signal from the sensor probe, converted it to ac, and then amplified it to a suitable level. The next step in this chain of events is ac to dc conversion. An electronic converter that

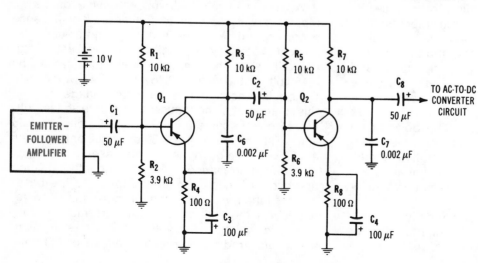

Fig. 7-9. Two-stage RC-coupled voltage amplifier circuit.

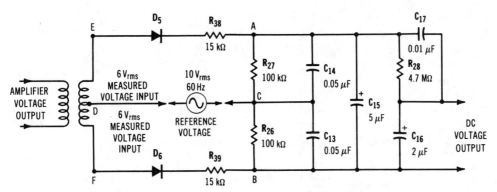

Fig. 7-10. Ac-to-dc phase-detector converter circuit.

achieves this operation is shown in Fig. 7-10. This circuit is commonly called a phase detector and has a number of applications in electronic instrumentation.

A phase detector is essentially composed of two diodes that are subjected to two ac signals. One of these is the ac signal from the voltage amplifiers. This signal is the measured component to which the circuit is to respond. The second signal is a reference voltage that is placed in series with the diodes. The resulting output of the circuit is a dc voltage that is dependent upon the phase relationship of the two input signals.

Three states of circuit operation will be discussed here. One of these occurs when no measured signal input is applied. The second two conditions occur when the measured signal is in phase or 180° out of phase with the reference signal. In practice, there are a number of phase variations between these two extremes.

A simplification of the phase detector converter is shown in Fig. 7-11 for the following explanation. Note that the transformer secondary winding is omitted in this circuit. In this case, we are assuming that no measuring signal is applied. Only the reference signal is applied to the circuit. This voltage is derived from a special power-supply transformer winding. Typical reference voltages are 10 volts rms, which produce a peak value of 14.14 volts and a peak-to-peak value of 28.28 volts.

With no measuring input signal applied to the phase detector, we can assume that only a reference voltage is applied to the diodes. Since the reference voltage is commonly fed between points D and C, both diodes will receive the same signal phase and voltage value at the same time. During the positive alternation, the diodes will be forward biased. Conduction occurs as indicated by the arrows and develops a voltage across R_{26} and R_{27}. No conduction occurs in either diode during the negative alternation. The direction of current in both circuits is determined by diode polarity. Current direction is, of course, from the cathode to the anode as indicated. This means that the current through R_{27} is from point C to point A in the circuit of diode D_5. The voltage at point A is, therefore, positive with respect to point C. Since D_5 offers very little resistance while conducting, the output will be the peak input voltage minus the diode voltage drop of 0.7 volt, or 13.44 volts. Diode D_6 produces an output voltage across R_{26}. The current through R_{26} is from point C to point B. Point B, therefore, is positive with respect to point C. The peak voltage of this output is the same as that across R_{27}. Referring back to Fig. 7-10, notice that capacitor C_{14} will charge to the peak voltage across R_{27}. Its discharge time is quite long, so for all practical purposes we will assume that the charge across C_{14} is 13.44 volts. Capacitor C_{13} will charge to the voltage across

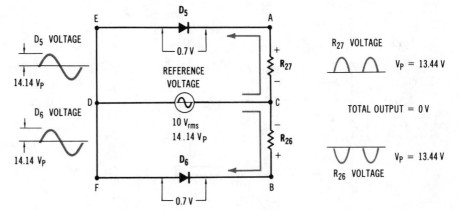

Fig. 7-11. Phase detector with no input signal.

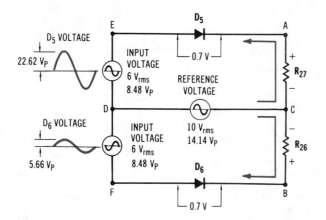

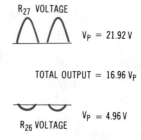

Fig. 7-12. Phase detector with in-phase input to D₅.

R_{26}. Capacitor C_{15} will then charge to the sum of these two voltages. Since the difference of potential between these two points (A and B) is 0 volts, C_{15} will have no charge.

The phase detector with a signal applied to the input is shown in Fig. 7-12. Notice that the transformer input between points E and D is in phase with the reference voltage. This, of course, means that the voltage from point D to point F is 180° out of phase with the reference voltage. An arbitrary measured voltage of 6 volts rms, or 8.48 volts peak, has been used for purposes of illustration. All voltages shown are peak values. The voltage applied to D_5 is between points E and C, which means the vector sum of the reference voltage and one-half the transformer input, since they are in series. These two voltages are in phase, so their vector sum is the direct sum of the peak voltages. In this case, it is 14.14 volts plus 8.48 volts or 22.62 volts peak. As mentioned previously, the diode output is developed across R_{27}. Its peak output voltage is 22.62 volts minus the 0.7-volt diode drop, or 21.92 volts. This voltage will charge C_{14} of the original circuit.

The reference voltage of 14.14 volts peak and the measured input voltage from points D to F of 8.48 volts peak are applied to diode D_6. Since these voltages are in series, they will add vectorially. In this part of the circuit, however, the

voltages are 180° out of phase. The resulting voltage will, therefore, be the vector sum, which will actually be the difference in the two voltages. This means a diode voltage of 14.14 volts peak plus (−8.4) volts peak, or 5.66 volts peak. With a diode voltage drop of 0.7 volt, the output across R_{26} will be 4.96 volts peak. A rectified output with this amplitude is developed across R_{26}.

In Fig. 7-10, you will recall that C_{13} will charge to the voltage developed across R_{26}. This voltage, in addition to that developed across C_{14}, is used to charge C_{15} to the total potential difference developed across points A and B. This would amount to 21.92 volts across C_{14} plus (−4.96) volts across C_{13}. Since these voltages are opposing, the total charge on C_{15} will be 21.92 − 4.96 or 16.96 volts peak. This essentially means that the ac voltage values applied to the phase detector will produce a corresponding value of dc voltage.

When the phase of the measuring input signal is reversed, the process changes as indicated in Fig. 7-13. The voltage between points E and D is now 180° out of phase with the reference voltage. A peak voltage of 5.66 volts is applied to diode D_5. The output across R_{27} is approximately 4.96 volts. C_{14} will, therefore, charge to this value. At the same time, the reference voltage and the voltage from point D to point F are in phase. This causes a combined voltage of 8.48 + 14.14, or

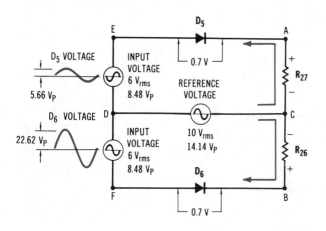

Fig. 7-13. Phase detector with in-phase input to D₆.

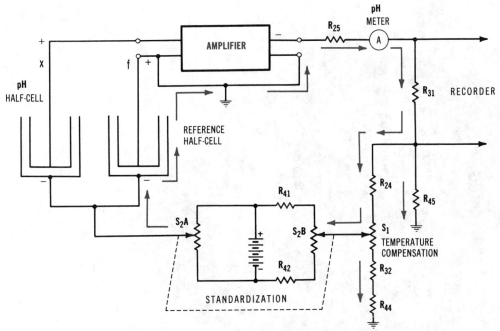

Fig. 7-14. Simplified schematic of amplifier feedback loop.

22.62 volts peak to be applied to D_6. The rectified output across R_{26} has a peak value of 21.92 volts. C_{15} will now charge to the combined peak voltages across C_{13} and C_{14}. The total charge appearing across C_{15} is 4.96 volts due to C_{14} plus (-21.92) volts due to C_{13} or -16.96 volts (point A with respect to point B). This essentially means that the second half of the measured input ac voltage produces a dc voltage value that is equivalent to the first alternation of the input but of opposite polarity.

In normal operation, the value of the measured input voltage changes according to the pH level of the solution being tested. As a result of phase detector operation, a dc voltage is developed that can effectively be used to drive a recorder or that can be measured by a meter to indicate specific pH values.

FEEDBACK LOOP

It has been shown that a positive input produces a negative output from the amplifier. Fig. 7-14 is a simplified schematic of the feedback loop. The negative output of the amplifier causes a current to conduct through R_{25}, the meter, R_{31}, R_{24}, S_1, R_{32}, and R_{44} in the direction shown. The amount of current depends on the output of the amplifier. This current is registered on the pH meter. The voltage drop across R_{31} also produces an output proportional to the feedback current for a recorder indication. S_1 is the adjustment for manual temperature compensation. If automatic temperature compensation is used, S_1 and R_{32} are switched

out of the circuit and a thermistor is switched in. This switching is accomplished by an auto-manual switch. The thermistor automatically changes resistance with temperature variations.

A negative voltage is developed at the wiper arm of potentiometer S_1. This voltage determines the operating voltages of the standardization bridge. The battery indicates the voltage provided by the regulated power supply. S_2 is set with a standard buffer solution. This determines the operating curve of the amplifier. If the potentiometer were set at the center position, the voltage across the bridge would be zero. The voltage at wiper arm A would be the same as at wiper arm B. S_2 can be adjusted so that point A can be slightly more positive or more negative than point B. The negative voltage at point A produces a current through the reference half-cell of the electrode. This produces a voltage drop that causes the circuit common to go more positive, which subtracts from the input to the amplifier. This is, of course, negative feedback.

PORTABLE pH INSTRUMENTS

Portable pH analyzers of the hand-deflection type have rather widespread usage in industry today. In general, these instruments utilize the basic circuitry discussed in this chapter. Battery power supplies with rechargeable power packs are typical with this instrument. The electrodes and standardizing material are conveniently stored in a hinged cover compartment. Fig. 7-15 shows a representative portable pH analyzer unit.

Fig. 7-15. A portable pH analyzer.

Courtesy Leeds & Northrup Co.

SUMMARY

The measurement of pH determines if a solution is either acid or alkaline (base). Specifically this refers to the number of positive hydrogen ions or negative hydroxyl ions in a solution.

The pH level of a solution is measured by the amount of dc voltage developed between two electrodes immersed in the solution under test. On a scale of 0 to 14, 7 is an indication of a neutral solution. The numbers from 7 down to 0 represent acid levels, while the numbers from 7 to 14 represent base levels.

The probe of a pH instrument is often called a half-cell or a cell. It develops a dc voltage when placed in a solution. A measuring half-cell and a reference half-cell may be combined in one probe or may be independent.

A pH instrument employs a measuring circuit, a standard circuit, dc-to-ac conversion, emitter-follower amplification, voltage amplifiers, phase-detector converters, and temperature compensation.

Negative feedback is used to minimize gain variations in the measuring circuit. This is only accomplished when amplifier gain is high.

When ICs are used, the power supply usually develops a split dc voltage. Typical values are +15 volts and −15 volts with respect to common or ground.

In the dc-to-ac converter, if a chopper is used its contacts should be normally open to keep the input impedance high.

An emitter follower is used to match the output impedance of the chopper or converter to the low input impedance of the multistage amplifier. An emitter follower is essentially a common-collector amplifier. The output is developed across a load resistor connected to the emitter. This resistor is of a small value, which determines the low output impedance of the amplifier. The power gain may be 100 with a voltage gain of less than one.

High levels of voltage amplification are needed to drive the output indicator. A multistage voltage amplifier achieves this function. RC-coupled amplifiers are commonly used to achieve this function in a pH indicator.

The next step in the chain of events is ac-to-dc conversion. An electronic converter called the phase detector is used to accomplish this operation. With only a reference signal applied, no output is developed by the detector. With a reference voltage and a measured signal voltage applied, an output is developed depending on the polarity of the measured input with respect to the reference signal.

CHAPTER 8

Beckman pH Analyzers

INTRODUCTION

Beckman Instruments makes a number of pH analyzers that are commonly used in industrial process applications today. These instruments accept an input signal from the pH sensor, which may be submersed in the solution under test or placed in a flow assembly. The developed measuring and reference signals are applied to individual high-impedance constant-gain operational amplifiers. The outputs from the two amplifiers are then applied to a differential amplifier, which is used to cancel solution potentials and produce an output that is proportional to pH only. This output is then applied to a variable-gain amplifier that has automatic temperature compensation. Ultimately the signal may be used to drive a recorder or to indicate pH level values on a hand-deflection meter.

In this chapter, pH analyzers will be divided into three sections that involve circuitry not previously discussed. The first section will be directed toward a transistor regulated power supply. The second part describes the circuitry of the preamplifier/differential-amplifier assembly that is used in the pH input circuitry. In addition to this, a dc-amplifier/current-output circuit will be discussed.

POWER SUPPLY

Fig. 8-1 shows a simplified schematic diagram of the master power supply used in a Beckman pH analyzer. This particular power supply uses a bridge rectifier to achieve a split dc voltage of +15 volts and −15 volts with respect to the common or ground connection. The split 15-volt outputs of the supply are used to energize ICs and discrete transistors in the composite circuit.

With 120 volts 60-Hz ac applied to the primary winding of the transformer, 25 volts ac appears across each half of the secondary winding. Assume now that the top of the secondary winding is positive and the bottom is negative during the first alternation. When this occurs, current conducts from the negative side of the secondary winding through D_2, Z_2, R_{L2}, R_{L1}, Z_1, D_4, and back to the positive side of the secondary winding. This alternation is indicated by solid arrows in the diagram.

The next alternation of the ac input causes the top of the secondary winding to be negative and the bottom to be positive. When this occurs, current conducts from the negative side of the winding through D_1, Z_2, R_{L2}, R_{L1}, Z_1, D_3, and back to the positive side of the secondary winding. This al-

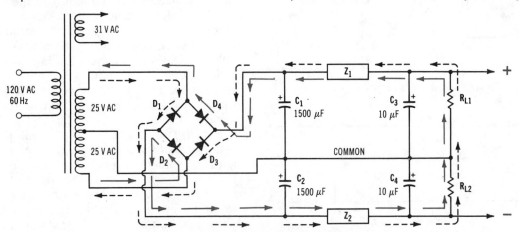

Fig. 8-1. Simplified schematic of a split power supply.

ternation is indicated by broken arrows in the diagram.

As a result of the current conduction just described, full-wave rectification is achieved by the bridge circuit. After filtering, the dc output will appear as 30 volts across R_{L1} and R_{L2}. If the common connection of these two resistors is connected to ground and returned to the center-tap connection of the transformer, the voltage is divided and becomes +15 volts and −15 volts with respect to ground. Resistors R_{L1} and R_{L2} are only used in this circuit to represent the load applied to the power supply. In practice, they are not specific components of the power supply. The positive half of the power supply is filtered by C_1, C_3, and Z_1. A duplicate filter is provided by C_2, C_4, and Z_2 for the negative half of the power supply.

TRANSISTOR VOLTAGE REGULATOR

Fig. 8-2A shows a simplified schematic diagram of a series-connected transistor voltage regulator. Two of these regulators are used in the split power supply just discussed. In Fig. 8-1, Z_1 and Z_2 indicate the location of these regulators in the circuit. We will describe the operation of the regulator first, then show it placed in the split power supply. Through this approach, you should have a better idea of how transistor regulators are used to achieve improved power supply control.

Power supply voltage regulation can be greatly improved through the use of transistors. The series transistor regulator, in this case, behaves somewhat like a simple series-connected variable resistor whose resistance is determined by circuit operating conditions. Fig. 8-2B is used to demonstrate the basic principle of transistor regulation.

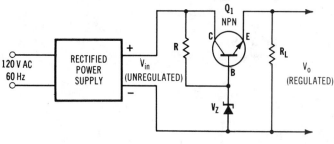

(A) Schematic diagram.

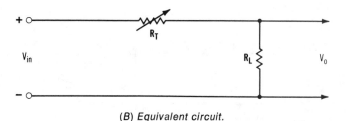

(B) Equivalent circuit.

Fig. 8-2. A series-connected transistor voltage regulator.

Assume now that an unregulated dc input is applied to the variable resistor circuit. This voltage is labeled V_{in}. The variable resistor is R_T and the power supply load is represented by R_L. The voltage developed by the load appears as V_o. For variations in R_L, if V_o is to remain at a constant value, the ratio of R_L to R_T must remain at a fixed level. This means that a change in load resistance, R_L, is compensated for by a similar change in the value of R_T.

Specifically, an increase in the resistance of R_L would cause a corresponding rise in V_o. To compensate for this, R_T should increase an equal amount. This, in turn, would lower V_o by causing an additional voltage drop across R_T. A decrease in the resistance of R_L would normally cause a corresponding decrease in the value of V_o. To compensate for this, R_T should be decreased an equal amount. This, in turn, would cause less voltage drop across R_T which would increase V_o accordingly.

The transistor regulator of Fig. 8-2A achieves control of V_o by changing its conduction capabilities according to voltage variations. Since the voltage across the zener diode is fixed, a decrease in V_o will result in a corresponding increase in the emitter-base voltage (V_{BE}) of Q_1. In this circuit, V_{BE} is determined by zener diode voltage V_Z minus V_o.

Assume now that a transistor regulator is placed in a circuit with an output voltage of 15 volts. If V_Z is rated at a fixed value of 18 volts, then V_{BE} would be 3 volts normally. Now assume that the resistance of the load increases in value, which causes V_o to rise to 17 volts. As a result of this action, V_{BE} will decrease because 18 volts (V_Z) − 17 volts (V_o) = 1 volt (V_{BE}). A decrease in V_{BE} will reduce the forward biasing of the transistor and cause it to increase in resistance. This, in turn, causes more voltage drop across the transistor, which returns V_o to its original 15 volts.

In the same manner, a decrease in V_o is commonly caused by a reduction in the resistance of R_L. If V_o, for example, decreases to 13 volts, it would cause a corresponding increase in V_{BE}. In this case, 18 volts (V_Z) − 13 volts (V_o) produces 5 volts (V_{BE}). With this increase in V_{BE}, the transistor will conduct more and have a reduced internal resistance. This, in turn, will produce less voltage drop across the transistor and increase V_o to its original 15 volts.

Fig. 8-3 shows a ±15-volt split power supply with transistor voltage regulation. Note that a regulator is placed in each half of the supply. The transistors must be of the opposite polarity in order to develop the correct output polarity. Transistor Q_1 is an npn, while transistor Q_2 is a pnp. The operation of the pnp regulator is primarily the same as the npn circuit just described. Through

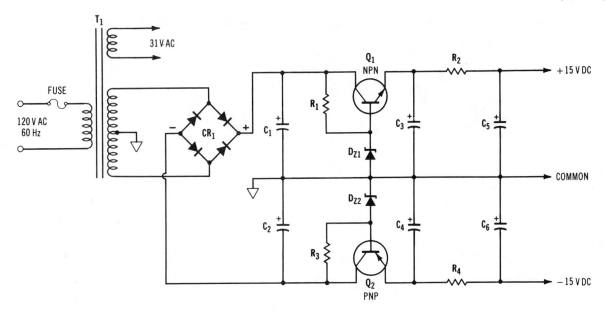

Fig. 8-3. Split power supply with transistor voltage regulation.

series transistor regulator circuits, improved power supply regulation can be obtained. In some circuits today the entire transistor regulator is built on a single IC chip. This obviously simplifies the circuit and its physical construction.

DIFFERENTIAL INPUT PREAMPLIFIER

In pH measurement, an unusual problem exists where any extraneous leakage currents to ground find a return path through the low resistance reference electrode. With a single input amplifier, this extraneous voltage adds to the pH potential resulting in a significant error in the determined pH level. When the resistance of the reference electrode is relatively low, the amount of error is insignificant. Should there be an increase in electrode resistance due to coating or junction clogging, the voltage increases a great deal, which produces a significant error in the indicated pH level.

In Beckman pH instruments, extraneous leakage currents are reduced by connecting a preamplifier in both the pH sensor and reference electrode inputs. With this type of input a negligible amount of current conducts through the electrodes. The glass electrode only measures pH potential through the solution to ground while the reference electrode measures the voltage through the solution to ground.

The outputs of the preamplifiers are then applied to a differential amplifier circuit. The function of this amplifier is to cancel solution potentials and produce a signal that is equal to only the pH potential. The simplified block diagram of Fig. 8-4 shows the preamplifier/differential-amplifier unit. In practice, the entire circuit assembly

is built on a printed-circuit board and built into the top of the electrode assembly. Fig. 8-5 shows a view of the assembly.

Signals from each electrode are applied to the noninverting input of a separate, high-gain op amp. With op-amp circuitry of this type, the input sees a high impedance. After high-level gain has been performed, each output is applied to the input of the differential amplifier. The pH signal is applied to the inverting input and the reference electrode signal is applied to the noninverting input. The output is thus representative of signal difference. The difference signal output is again amplified by a high-gain op amp. In this case, the amplifier is temperature compensated by a sampling signal from the solution under test. In practice, the temperature compensation signal is derived from an electrode that is in contact with the pH sample electrode.

A number of unique advantages in pH measurement are present in the Beckman circuit. This includes elimination of drift and noise, reduced maintenance problems, remote pH instrument placement, and elimination of costly interconnecting coaxial cables.

DC AMPLIFICATION/CURRENT OUTPUT

Fig. 8-6 is a schematic diagram showing the dc-amplifier/current-output circuitry of a Beckman pH analyzer. In this circuit, the dc output signal of the preamplifier/differential amplifier is applied to the input of the circuit at terminal 3 of terminal strip P_1. Dc amplification, a comparator amplifier, dc to dc inversion, and current output are all included in this part of the pH analyzer. The current output is ultimately used to indicate pH levels on

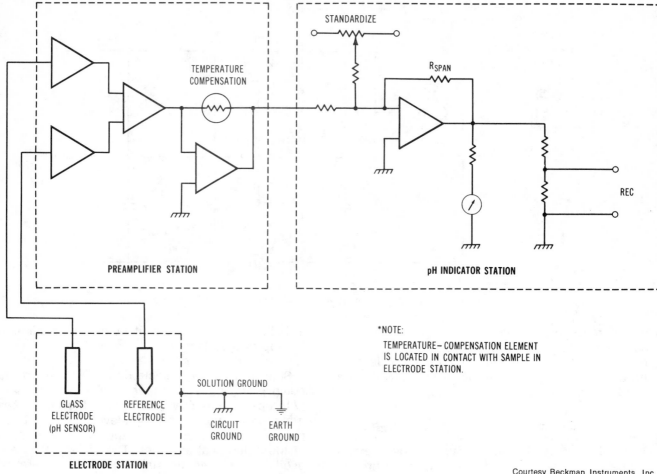

Fig. 8-4. Simplified block digram of a Beckman pH analyzer.

Courtesy Beckman Instruments, Inc.

Fig. 8-5. Differential amplifier assembly of a Beckman pH analyzer.

Courtesy Beckman Instruments, Inc.

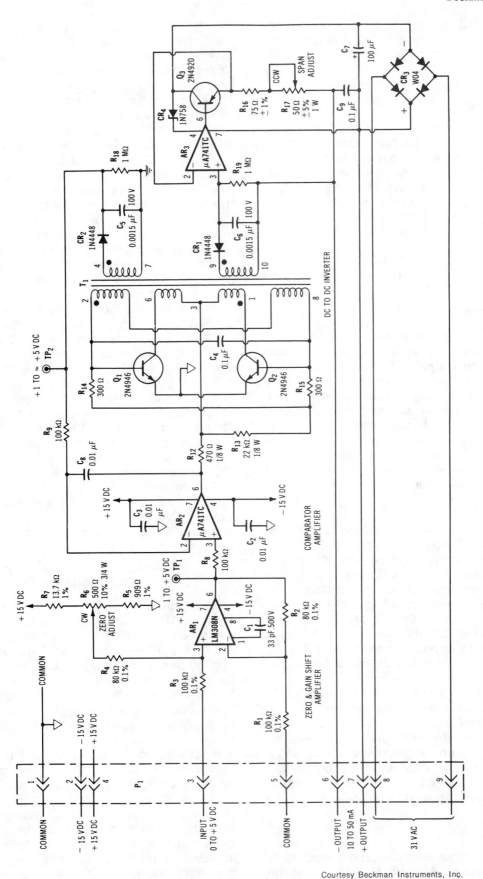

Fig. 8-6. Schematic diagram of the current output module of a Beckman pH analyzer.

a hand-deflecting instrument or to drive a chart recorder. The entire circuit is built on a printed-circuit board and is called the current output module.

DC Amplification/Zero Adjust

Op amp AR_1 is a dc amplifier that has the dc signal applied to its noninverting input. Potentiometer R_6 is the zero adjust control used to compensate for circuit voltage variations that are generally due to input loading. Resistors R_1 and R_2 determine the total gain of the amplifier. Voltage amplification (A_v) is based on the ratio of R_1 to R_2 and is determined by the formula

$$A_v = \frac{R_1 + R_2}{R_1}$$

Comparator Amplifier

Op amp AR_2 is a comparator amplifier. It essentially compares the voltage levels of signals applied to its two inputs. The dc pH signal is applied to the noninverting input and a reference or feedback signal is applied to the inverting input. This op amp simply compares the measured pH signal with the reference and produces an output. If the pH measured input is larger than the reference, a positive output will appear at pin 6. If the input is smaller in value than the reference, a negative output will occur.

DC to DC Inverter

The term *inverter* is a common way of describing the primary function of components Q_1, Q_2, T_1, CR_1, and CR_2. In general terms, dc is first changed to ac by an oscillator composed of Q_1 and Q_2. The ac signal is then stepped up by transformer T_1. The two secondary winding voltages of T_1 are then rectified and filtered by CR_1-C_6 and CR_2-C_5, respectively. The resulting dc signals are either used to form a feedback signal, as in the case of CR_2-C_5, or to drive the output circuit, as achieved by CR_1-C_6.

The operation of a circuit similar to the dc to dc inverter was discussed in conjunction with the power supply of Chapter 6. The inverter of Fig. 8-6 is similar in operation to that of the previous circuit with the primary differences being input voltage values and application of the output circuitry. The circuit is called a dc to dc inverter here and was described as a dc to ac converter in Chapter 6.

Current Output

The current output section of the module is achieved by op amp AR_3, transistor Q_3, and the bridge rectifier CR_3. The circuit indicated has a 10- to 50-milliampere output capability. The current span is adjusted by R_{17}.

Negative dc voltages are developed by CR_1, filtered by C_6, and applied to the noninverting input of op amp AR_3. The op amp responds in this case as a noninverting amplifier. The negative dc output voltages appearing at pin 6 of AR_3 are connected directly to the base of pnp transistor Q_3. These voltages forward bias the base-emitter junction of Q_3. The collector-base junction of Q_3 is reverse biased by the dc output voltage of bridge rectifier CR_3. More negative base voltage values produce increased current output, while less negative base voltages produce decreased current output. The span adjusting control, R_{17}, alters the emitter current conduction level of Q_3. When R_{17} is reduced in resistance, the current span is raised; when R_{17} is increased in resistance, the current span is lowered.

The bridge rectifier, CR_3, serves as a full-wave, high-current source for the output circuit. An ac voltage of 31 volts from the main power transformer is supplied to the bridge through external connecting wires at terminals 8 and 9 of P_1. The bridge then rectifies the ac and filters it with C_7. Approximately 27 volts of dc is applied to Q_3. With respect to the common ground, the dc output of CR_3 is approximately +13.5 volts and −13.5 volts. This voltage is also used to supply op amp AR_3. Zener diode CR_4 is used to regulate the −13.5-volt side of the source.

The current output of the pH analyzer may be used to drive an indicating meter or a recorder. Fig. 8-7 shows a strip chart recorder pH analyzer, and Fig. 8-8 shows a pH analyzer with a hand-deflection indicating meter.

SUMMARY

Beckman pH instruments are commonly used in industrial process applications today. In these instruments the measuring and reference signals are both applied to op amps. Each output is then applied to a differential amplifier which is used to cancel solution potentials. A variable-gain amplifier is then used for temperature compensation. The output is ultimately used to drive a recorder or to indicate pH values on a hand-deflection meter.

A split dc power supply is used to provide the operating voltages for the pH indicator. In this circuit, a bridge rectifier is used with series transistor regulators. The series transistor regulator responds as a variable resistor that changes value according to circuit operating conditions. An increase in the value of R_L is compensated by an increase in transistor resistance, which maintains the output voltage at a constant value. A decrease in R_L would normally cause a drop in V_o. To com-

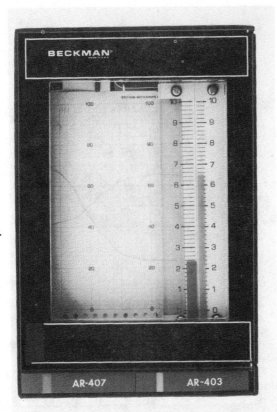

(*A*) *Front view of instrument.*

(*B*) *Partially removed from housing.*

Fig. 8-7. A Beckman strip chart recorder pH analyzer.

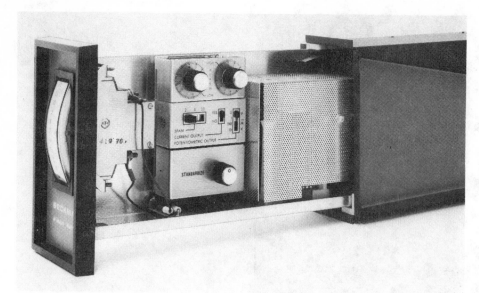

Fig. 8-8. A Beckman pH analyzer with hand-deflection indicating meter.

Courtesy Beckman Instruments, Inc.

pensate for this, conduction of the series transistor is increased to reduce its series resistance. As a result, V_o is increased. Through variations in transistor resistance the output voltage is maintained at a constant value.

In the differential input preamplifier, individual op amps are used in each of the two inputs. These high-impedance op-amp inputs are used to reduce extraneous leakage voltages to ground. The output is then applied to a differential amplifier. The pH electrode signal goes to the inverting input and the reference electrode signal goes to the noninverting input. The signal difference is then amplified to reduce drift, noise, and temperature variations.

An op-amp dc amplifier is connected to the output of the differential amplifier. The gain of this amplifier is determined by the formula $A_v = (R_1 + R_2)/R_1$.

The comparator amplifier is used to determine the levels of signals applied to its two inputs. The dc pH signal is applied to the noninverting input and a reference or feedback signal is applied to the inverting input. A comparison of the two signals is made so that when the pH signal is larger a positive output will occur. If the feedback/reference signal is greater a negative output will occur.

The pH instrument of this chapter employs a dc to dc inverter. Essentially this is similar to the dc to ac converter discussed previously.

The current output is achieved by an op amp and a power transistor. An output of 10 to 50 milliamperes is typical for this instrument.

CHAPTER 9

Magnetic Flowmeters/Converters

INTRODUCTION

A cutaway view of a Fischer & Porter magnetic flowmeter is shown in Fig. 9-1. This type of unit is a volumetric fluid flow-rate detector that changes conductive fluid flow into an induced voltage when the fluid flows through a magnetic field. The amplitude of the generated signal is directly proportional to the flow rate of the fluid.

Magnetic flowmeters utilize obstructionless metering in their operation. An inherent advantage of this principle is that pressure losses are reduced to levels occurring in equivalent lengths of equal diameter piping. This essentially reduces and conserves pressure source requirements compared with other metering methods.

Fig. 9-2 is a block diagram of a Fischer & Porter magnetic flowmeter and solid-state converter to

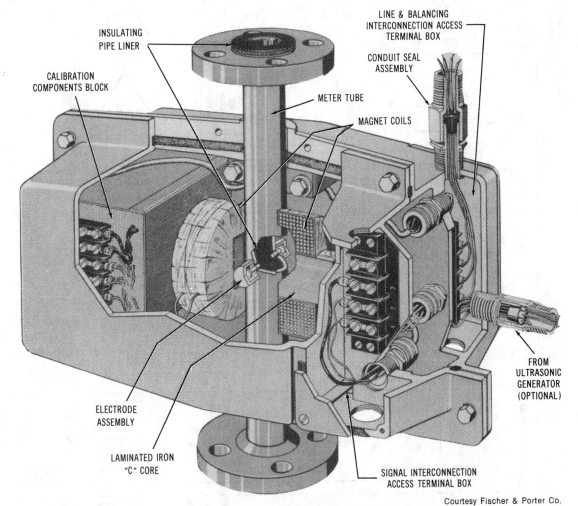

Courtesy Fischer & Porter Co.

Fig. 9-1. Cutaway view of a magnetic flowmeter.

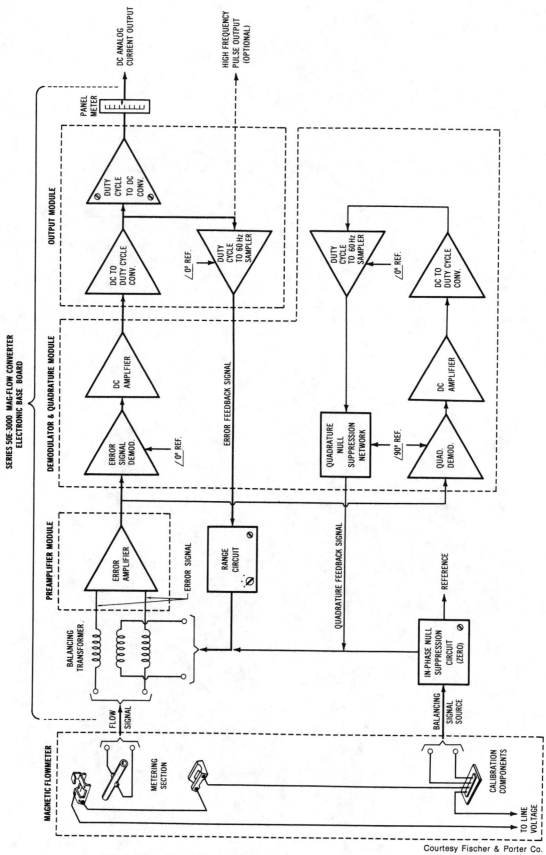

Courtesy Fischer & Porter Co.

Fig. 9-2. Functional block diagram of a magnetic flowmeter/converter.

be discussed in this chapter. The discussion of this unit is divided into five major sections. This includes the power supply, magnetic flowmeter, input circuits, error signal loop, and the suppression signals. A large part of the electronics circuitry of this type of instrument is built on printed-circuit boards or cards that can be exchanged or replaced should any problem occur. Instead of the discrete component circuits described in this discussion, a large part of the circuitry is achieved by integrated circuits today.

POWER SUPPLY

The power supply of a magnetic flowmeter/converter is primarily responsible for delivering ac and dc power to the flowmeter and all of its active components. Two dc power supplies are utilized in the Fischer & Porter instrument. One of these is a full-wave rectifier with an RC filter that supplies −45 volts dc to the power amplifier circuitry. In addition to this, a full-wave bridge rectifier with a center-tapped transformer is used to develop +15 volts dc and −15 volts dc with respect to a common return. This is used to supply dc to the integrated circuits and solid-state components.

AC Power Distribution

The input line voltage of a magnetic flowmeter/converter is normally 117 volts ac, 60 Hz. This voltage is connected first to the primary winding of the power transformer of the converter unit that serves as the ac source for both dc power supplies. The transformer has a single primary and two secondaries that operate independently.

The 117-volt ac 60-Hz line voltage is also supplied to the magnetic flowmeter and calibration components. This voltage is connected in parallel with the converter unit. The 60-Hz line frequency also serves as a reference sampling source for the feedback circuit to be discussed later.

−45 Volts DC Power Supply

The −45-volt dc power supply of Fig. 9-3 is a simple full-wave rectifier using two diodes. The

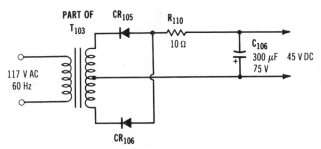

Fig. 9-3. Schmatic diagram of a full-wave −45-volt dc power supply.

filter is an RC network using a 10-ohm resistor and a 300-microfarad capacitor. The output voltage of this power supply is used to energize the power amplifier module circuit board.

+15 Volts DC and −15 Volts DC Power Supply

Voltages of +15 volts dc and −15 volts dc with respect to the common return are provided by the full-wave bridge power supply of Fig. 9-4. Each half of the secondary with respect to the center-tap connection provides 30 volts ac to the bridge diodes. The diodes are rated at 1 ampere with a 200-volt peak reverse voltage (prv) rating.

Each half of the power supply is filtered by a pi network with respect to the common return. The polarity of the capacitors of each filter is connected to accommodate the correct output voltage polarity. The two zener diodes regulate the output voltages to a constant +15 volts and −15 volts. This power supply furnishes voltage to all of the printed-circuit boards.

MAGNETIC FLOWMETER

Fig 9-5A is a schematic representation of the magnetic flowmeter, while Fig. 9-5B is a pictorial representation of the basic operating principles involved. The flowmeter is constructed around a section of pipe that requires no orifices or obstructions within it. This means that the flowmeter itself has very little effect on the flow rate of the fluid through it. Around the metering section of pipe are wound two field coils; one above the piping,

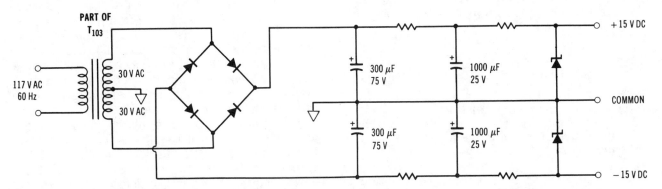

Fig. 9-4. Schematic diagram of a full-wave bridge ±15-volt split dc power supply.

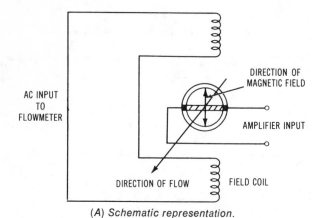

(A) Schematic representation.

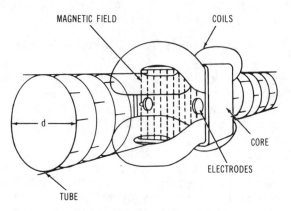

(B) Pictorial representation.

Fig. 9-5. Magnetic flowmeter.

the other below it. When current is passed through these coils, a magnetic field is produced in the direction shown in Fig. 9-5. The flowmeter section of the pipe has two electrodes positioned so that the fluid passing between them is perpendicular to the magnetic field of the coils. These two electrodes are electrically insulated from the walls of the tubing.

We will first consider the operation of the flowmeter when a constant magnetic field is applied. Under these conditions, the operation of the flowmeter is based on Faraday's law of electromagnetic induction. Simply stated, the voltage induced across a conductor moving at right angles through a magnetic field is proportional to the velocity of that conductor, or

$$V = \frac{1}{c}Bdv$$

where,
    c is a dimensional constant,
    B is the flux density of the magnetic field,
    d is the inside diameter of the pipe,
    v is the velocity of the conductor.

The conductor in this case is that element of fluid that lies directly between the two electrodes. As fluid passes through the pipes, it is moving perpendicularly through the magnetic field B. The value of B depends on the amount of current through the windings of the field coil. For the present, this will be considered as a constant. The inside diameter of the pipe is, of course, also a constant. For most liquids with a reasonable conductivity, the constant c is considered to be 1. As a´ result, V is dependent only on the velocity of the conductor and is directly proportional to it. The induced voltage V is the flowmeter output, which is used as the amplifier input. This input is directly related to the velocity of the fluid through the flowmeter.

The magnetic field B, however, is not constant. An ac voltage is applied to the field coils. This means simply that the output voltage will be an ac voltage rather than a dc voltage. Its magnitude will still be proportional to the velocity of the fluid. The desired signal output of the flowmeter will also be in phase with the magnetic field current. That is, when B is maximum, maximum voltage is induced. B is maximum when the current through the field coils is maximum.

Since an ac voltage produces the magnetic field, some unwanted voltage will be produced when there is no flow. When there is fluid in the flowmeter and no flow, a stationary conductor is placed between the two electrodes. A voltage will be produced between these two electrodes, due to the changing field. In other words, this voltage is produced by transformer action. This voltage is maximum when a maximum change in the magnetic field occurs. Maximum change in the field is produced by maximum change in current, which occurs in an ac circuit when the current is passing through zero. Therefore, this unwanted signal is 90° out of phase with the desired signal. The feedback circuit will provide for elimination of this signal.

Another unwanted signal is also produced by transformer action. This voltage will be induced into the electrodes and their leads. These leads are formed into a twisted pair and shielded where possible to minimize these unwanted signals. They must, however, be exposed near the electrodes and cannot be formed in a twisted pair around the pipe. This voltage must also be 90° out of phase with the desired signal. All signals 90° out of phase with the desired signal are called quadrature voltages. They will be canceled out in the feedback circuits so that the only amplified signals will be proportional to the flow rate.

A zero-flow-rate signal is also developed in the feedback circuitry. This signal will compensate for any in-phase voltages fed into the amplifier at a zero flow rate. It will also compensate for any change in the magnetic field of the flowmeter due to changes in line voltages.

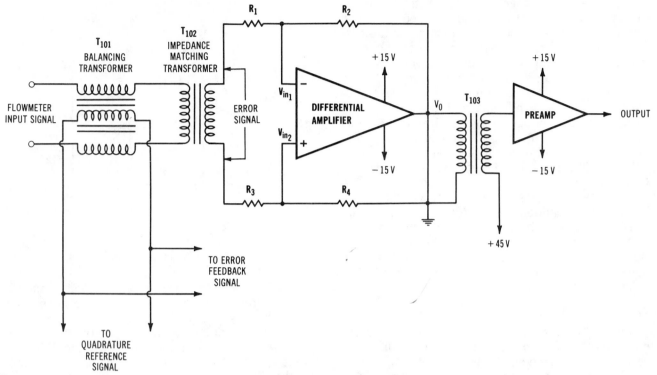

Fig. 9-6. Differential amplifier.

DIFFERENTIAL AMPLIFIER

Fig. 9-6 is a schematic diagram of a differential amplifier that is representative of the one used in a flowmeter converter. In this circuit, small input signals that are often buried in noise can be recovered and amplified above the noise. A high-gain operational amplifier is commonly used to achieve this function.

When a differential amplifier has signals of equal amplitude and polarity applied to each of its inputs at the same time, the output will be zero. In this case, the gain is

$$A_V = \frac{V_o}{V_{in_1} - V_{in_2}}$$

This is called the common-mode rejection characteristic of an op amp. If there is a difference in the two applied signals, this will be amplified and appear in the output above all other signals. In actual circuit operation, ac noise common to both inputs is rejected by the differential amplifier.

The high input impedance of the flowmeter is matched to the op amp through balancing transformer T_{101} and impedance matching transformer T_{102}. The output of the differential amplifier is developed across the primary of T_{103}. The induced secondary voltage of T_{103} serves as the input to the preamplifier stage. Feedback voltage is fed equally into the two inputs of the op amp through the primary winding or center coil of transformer T_{101}. Three voltage signals may be present at the

transformer primary. The first of these is the reference, or zero-flow-rate signal. When the instrument is properly adjusted to the zero-flow-rate point, this signal will appear at the primary or center winding of transformer T_{101}. The resulting induced signals will be in phase and at the same amplitude as the zero-flow-rate signal applied to the two secondary windings of T_{101}. The combination of these two signals produces a resulting zero input signal to the differential amplifier. In a like manner, the quadrature signal, which is 90° out of phase, is also canceled.

Reviewing the differential amplifier, notice that the only signal amplified by the differential amplifier is the error signal. This voltage is the difference between the flowmeter signal derived from the flow of fluid, and the reference signal. Quadrature and reference voltages are canceled by differential amplifier action. The output of the differential amplifier is then coupled to the preamplifier for further amplification.

VOLTAGE AND POWER AMPLIFICATION

Fig. 9-7 is a schematic diagram of a typical discrete component preamplifier. Discrete component preamplifiers are used in older converters, while IC op amps are used in the newer converters.

Preamplifier

In the circuit of Fig. 9-7, the input signal is transformer-coupled from the differential amplifier to the base of Q_2. Capacitor C_6 makes the input circuit resonant to 60 Hz. A large input 60-Hz signal can, therefore, be developed at the input to the preamplifier. Unwanted high-frequency noise signals will have a low-impedance path to ground through C_6. Emitter voltage is maintained across C_7 and R_{11}. The bias voltage is derived from the bias control (not shown). The bias voltage can be adjusted from 0 to about +15 volts. Resistor R_{10} acts as a load for Q_2. It is connected to the −30-volt power supply. With the proper bias, the collector voltage will be about −0.8 volt dc. The output of Q_2 is directly coupled to the base of Q_3. The collector voltage of Q_2 also biases Q_3. Diode D_1 maintains the emitter at a dc ground potential while preventing any high reverse current when a large signal is applied. Resistor R_{12} is the load for Q_3. Resistors R_{12} and R_{13} form a voltage divider that helps maintain the collector voltage at about −4.5 volts dc.

Phase Splitter

Fig. 9-8 is a schematic diagram of a discrete component phase splitter used in some solid-state converters. A phase-splitter is a circuit that is quite often used to provide two outputs that are 180° out of phase to drive a push-pull amplifier. It takes advantage of the fact that the emitter signal is in phase with the input while the collector signal is 180° out of phase with it. The only drawback in using this circuit is that, like the conventional emitter follower, its gain is always less than 1. The application of the circuit here, however, is somewhat unusual. Here it is used as a phase-setting, or phase-compensating, circuit.

The output of the preamplifier is RC-coupled through C_8 and R_{14} to the base of Q_4. Resistor R_{16} is the collector load resistor, while R_{17} is the emitter load. Bias is obtained by the voltage divider of R_{17} and R_{18}. The emitter voltage thus formed is about +2.5 volts dc. Instead of using two outputs as is usually done with this circuit, the output is

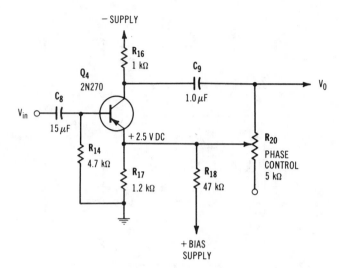

Fig. 9-8. Phase splitter.

taken from the collector. This output is applied to an RC circuit composed of C_9 and the phase control, R_{20}. By changing phase control R_{20}, the amount of resistance in the RC combination can be changed. You will remember that the RC relationship determines the phase shift across it. By varying R_{20}, the phase shift of the amplifier output can be changed from 20° to 120°. The adjustment will not only compensate for any phase shifts within the amplifiers, but will also give the proper phase relationship for the operation of the servomotor.

Voltage Amplifier

Fig. 9-9 is a schematic diagram of a typical voltage amplifier that is used in some solid-state converters. Newer converters generally achieve voltage amplification with IC op amps. In the discrete component amplifier you will notice that the circuits are almost identical to those in the preamplifier. The negative voltage supply is derived from the same point as for the preamplifier. The bias voltage is set by the bias control, as was true

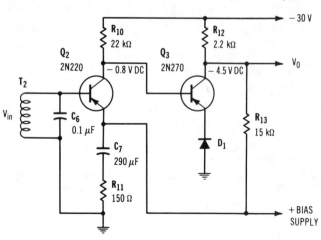

Fig. 9-7. Preamplifier.

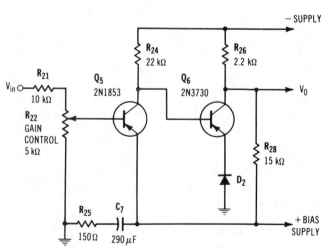

Fig. 9-9. Voltage amplifier.

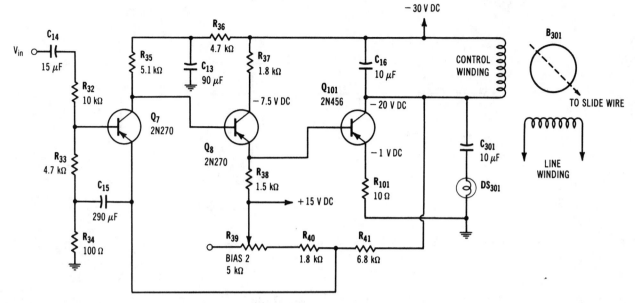

Fig. 9-10. Power amplifier.

in the preamplifier. We will refer to this bias control as the Bias 1 control. The only differences in the two circuits are found in the Q_5 circuit. First of all, Q_5 is a 2N1853 transistor rather than a 2N220. The base circuit contains a limiting resistor, R_{21}. The input signal is developed across R_{22}, which is the gain control of the amplifier. The output of the voltage amplifier is fed to the power amplifier.

Power Amplifier

Fig. 9-10 is the schematic diagram of the power amplifier. Transistor Q_7 is actually a voltage amplifier, but it was included in this section since all three of these circuits are directly coupled. The input to this circuit is RC-coupled to the base of Q_7 by C_{14} and the voltage divider made up of R_{32}, R_{33}, and R_{34}. The signal across R_{33} and R_{34} is the input to this stage. The bias for this stage comes from a voltage divider made up of R_{39}, R_{40}, R_{41}, and the balancing motor-control winding. A negative 30 volts is connected to one end of this voltage divider and a positive 15 volts at the opposite end. The Bias 2 control fixes the emitter voltage of Q_7. The voltage can be controlled from about -7.5 volts dc to $+7.5$ volts dc. C_{13} is a bypass capacitor. R_{35} is the load resistor for Q_7.

The circuit involving Q_8 is connected as an emitter follower. R_{38} is the load resistor. This resistor is connected to the $+15$-volt source. Bias is obtained by the dc voltage across this resistor. Resistor R_{37} in the collector circuit is used only to limit the collector current. The 45 volts across this transistor is very high for this type of transistor. If R_{37} were not in the circuit, an excessively high current might result. The output signal across R_{38} is directly coupled to the base of Q_{101}.

Transistor Q_{101} is a high-voltage transistor. The 2N456 is capable of handling about 40 watts. Resistor R_{101} develops the needed -1-volt bias due to emitter current through it. The control winding of the balancing motor is the load for the circuit. Capacitor C_{16} compensates for the inductance of the control winding so that the two together act as a resistive load. DS_{301} is an indicator lamp that provides for a visual indication of a signal regardless of phase. This helps to adjust zero and quadrature controls without using other measuring instruments. Capacitor C_{301} is a dc blocking capacitor, so only an ac signal can light the lamp.

FEEDBACK CIRCUITS

Fig. 9-11 is the schematic diagram of the feedback circuits. Terminals 1 and 2 are connected in series with the flowmeter field coils. The amount of voltage developed across these two points is determined by the combined resistance of the resistors between these two points. The trim adjustment, R_{45}, can vary this voltage. The voltage dividers, R_{42} and R_{43}, determine the zero reference point. The voltage at terminal 1, therefore, will be positive, while terminal 2 will be negative. The positive voltage will be three times as large as the negative voltage. R_{301} is the quadrature voltage adjustment whose position determines the amplitude and polarity of this voltage. C_{201} provides the 90° phase shift so that the quadrature feedback signal will be in phase with the quadrature flowmeter signals.

The amplitude and polarity of this signal is determined by the position of R_{304}, the zero adjustment. This signal will be in phase with the zeroing signal since the current through the feedback circuits is

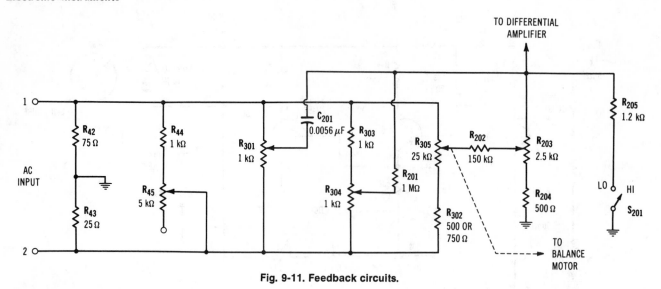

Fig. 9-11. Feedback circuits.

in phase with the flowmeter current. This was the reason for these circuits being connected in series.

The feedback voltage to the differential amplifier is developed across R_{203} and R_{204}. R_{203} is the range adjustment. The positioning of this potentiometer determines the effect of the slide-wire variation on the differential amplifier. A coarse range adjustment is provided by switch S_{201}. In the Lo position, the impedance across which the feedback voltage is developed is decreased. This decreases the voltage at this point and also limits the range of voltages to be fed back to the differential amplifier. R_{305} is the slide-wire potentiometer, which is positioned by the balancing motor. When an input is fed to the amplifier, the output drives the slide wire to a point where the feedback from R_{305} to R_{203} and, hence, to the differential amplifier, completely nullifies the input. The balancing motor also drives the flow-rate indicator.

SUMMARY

Magnetic flowmeters utilize obstructionless metering in their operation. Through this principle of operation, flow can be measured without pressure losses occurring in equivalent lengths of piping. This reduces pressure source requirements compared with other metering methods. The complete unit includes a flowmeter and a converter circuit.

The power supply of the flowmeter/converter discussed in this chapter employs a full-wave rectifier with an RC filter. The power amplifier of the circuit requires −45 volts dc, while the ICs and discrete components are supplied +15 volts and −15 volts dc with respect to ground. Line voltage of 117 volts ac, 60 Hz is supplied to the power supply transformer and to the flowmeter unit at the same time.

Flowmeter operation is based on Faraday's law, which states that the voltage induced across a

conductor moving at right angles through a magnetic field is proportional to the velocity of the conductor. If the conductor is a fluid, it can be used to develop a resulting output. The induced voltage is considered as the flowmeter output. It must be amplified and processed by the converter before it will be of value.

A differential amplifier is part of the converter circuitry. An amplifier of this type has input signals applied to both the inverting and noninverting inputs. The amplified output is the difference between the two input signals. If the input signals are equal in amplitude and of the same polarity, the output is zero. This is called common-mode rejection. The only signal amplified is the error signal, which is representative of the flow of fluid compared with the reference signal. The output of the differential amplifier is then applied to the preamplifier.

A preamplifier achieves high signal gain and it has good noise-rejection capabilities. High-frequency noise is passed into ground by capacitors. Two or more discrete component amplifiers are commonly used in the preamplifier of the flowmeter converter.

A phase splitter or phase shifter is used in the converter of the flowmeter to compensate for phase shift within amplifiers and to establish proper phase relationships for the servomotor connected to its output. Phase shifting is achieved by changing a phase control. Phase shifting of 20° to 120° is typical of this circuit.

The voltage amplifier of the converter is usually two or more discrete bipolar transistors. Normally these amplifiers are direct-coupled. A gain control is included in the amplifier input to compensate for different levels of signal input.

The power amplifier circuit employs two direct-coupled voltage amplifiers and a power amplifier. The first two transistors serve as signal drivers for the power amplifier. The power amplifier output is normally used to drive the servomotor or indicator assembly.

The converter of a flowmeter employs a feedback circuit that returns to the differential amplifier. A slide-wire potentiometer is positioned by the balancing motor. When an input signal is applied to the amplifier, the output drives the slide-wire resistor to a point where the feedback signal ultimately nullifies the input signal.

CHAPTER 10

Transmitters

INTRODUCTION

In this chapter we will discuss transmitters or transducers used in the Honeywell ElectriK Tel-O-Set. In order to do this we need to have an understanding of the type of oscillator used—a modified version of the Hartley oscillator. A basic version of the oscillator will be discussed first, then the modified version will be discussed pointing out the major differences in circuit design. A block diagram of the Honeywell unit appears in Fig. 10-1.

The oscillator controls the direct current shown here as the output. The sensing device used by the oscillator is a coil in the detector assembly. The inductance of this coil varies with the position of the pivoted beam. The beam position is determined by an input working against the spring that provides a method of mechanically zeroing the transmitter. Feedback repositions the pivoted beam by the magnet unit. The output of this transmitter can be used by several different types of receivers, indicators, or controllers. These components, as well as the power supply, will be covered in subsequent paragraphs.

Finally, the type of transducers available will be discussed. It will be shown how each of the inputs is transformed into usable outputs. The input can be pressure, current, process pressure, or differential pressure. The output is either current or pressure.

HARTLEY OSCILLATOR

Fig. 10-2 is a schematic diagram of a series-fed Hartley oscillator. The identifying characteristic of this oscillator is the tapped coil in the base-emitter circuit. Coil L_1 and capacitor C_1 form a resonant tank circuit. The values of these components determine the frequency at which the circuit oscillates. The base capacitor, C_B, and the base resistor, R_B, provide automatic or self bias for the oscillator. The emitter of the transistor is connected to the coil in the tank circuit. This circuit provides an output to input feedback path, which is necessary for any oscillator circuit to operate. All of the emitter current of the circuit must pass through this series part of the coil. As a result of this connection, the term series-fed oscillator is used to describe this circuit condition.

The Hartley oscillator is often found in radio receivers and transmitters. Oscillators are important in the field of electronics because they are high-frequency ac generators. Oscillators provide ac voltage at frequencies at which electromechanical generators are neither practical nor possible. Four circuit characteristics are necessary before an oscillator can function. It must:

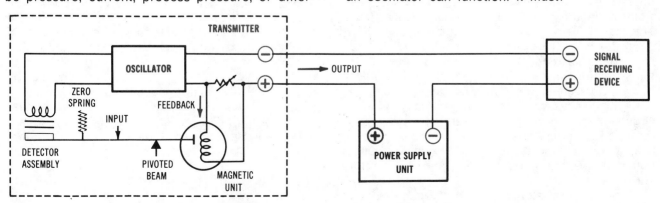

Fig. 10-1. Block diagram of Honeywell ElectriK Tel-O-Set.

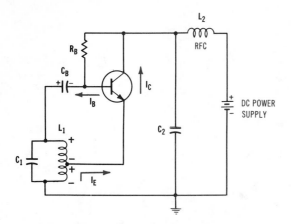

Fig. 10-2. A series-fed Hartley oscillator.

1. Be an amplifying device.
2. Have a source of dc power.
3. Have a frequency-determining device.
4. Have positive feedback.

The Hartley oscillator meets all of these requirements. When dc power is applied, the transistor starts to conduct because of its forward-biased base-emitter junction. The path of this current is from the negative side of the power supply, through the lower windings of coil L_1 to the tap, to the emitter, the collector, and back to the positive side of the power supply. As the current through the tapped portion of L_1 increases, a voltage is induced in the entire coil by transformer action. The lower portion of the coil acts as the primary of the transformer, whereas the entire coil is the secondary. When the primary and secondary of a transformer are a part of the same coil, it is called an autotransformer. The polarity of the voltage induced on the base of the transistor is such as to provide positive feedback or regeneration. As shown in the schematic, a rising current through the coil primary induces a positive voltage on the base side of the coil. This positive voltage is applied through C_B to the base of the transistor, which causes a rise in base current through the coil, thus inducing a more positive voltage on the base. The feedback cycle continues until the transistor collector-emitter current reaches saturation. This action takes place almost instantly.

As the base becomes more and more positive, more base current is applied to capacitor C_B in the direction shown. When the positive voltage across L_1 is removed because of transistor saturation, this leaves the base with a relatively high negative voltage from C_B. Then C_B discharges through R_B. During this time, the base is held negative, which reverse biases the base-emitter junction and cuts off the transistor collector-emitter current.

When the transistor reaches saturation, its collector-emitter current no longer rises. No trans-former action can occur at this time since the magnetic field remains constant. As a result, no voltage will be induced across L_1. The base voltage then begins to go negative due to the charge on C_B. The base current decreases as a result. A decreasing current through the primary of L_1 induces a negative potential at the base side of L_1. This negative cycle of operation continues during the cutoff time of the transistor, however, because of the "flywheel effect" of the LC resonant tank circuit. The oscillating process is thus continuous. The frequency of oscillation is determined by the values of L and C:

$$f_o = \frac{1}{2\pi\sqrt{LC}}$$

TRANSMITTER OSCILLATOR

Fig. 10-3 is a complete schematic diagram of the transmitter oscillator circuit. A dc supply of 42 volts is necessary for the proper operation of the oscillator. Current and voltage receiving units will be placed in series with the power supply. A variable series resistor adjusts the load current so it will have a value of between 4–20 milliamperes. The voltage drop across the 2.5-ohm resistor is used as a test voltage. The 80-microfarad capacitor filters out any ac voltage that might be present across it. A direct-current path is provided around the magnet unit by L_2, the 750-ohm resistor, and the coarse and fine span adjustments which determine the amount of feedback fed to the balance beam by the magnet unit. The 250-microfarad capacitor is a bypass capacitor. These componets, with the exception of the magnet unit, have the primary function of controlling the current (or voltage) output.

Let us now consider the operation of the oscillator section of Fig. 10-3. Its operation is essentially the same as that of the Hartley oscillator previously discussed. One of the first noticeable differences is that it uses two transistors instead of one. In this case, Q_1 and Q_2 are connected in parallel to provide sufficient output current. Resistors R_1, R_2, and R_3, and diode D_1 provide the voltage drop that determines the emitter potential. Diode D_1 has the added responsibility of sensing the amplitude of the base signal and providing for changes in the forward bias of the transistors that are reflected in current output. Base voltage is determined by the voltage divider formed by R_5, R_6, R_7, R_8, and R_9. Both R_7 and R_9 are temperature-compensating resistors. L_1 and C_2 make up the Hartley resonant circuit and determine the frequency of operation. The feedback path is through C_1, the tapped portion of L_1, and the detector coil, which senses the position of the pivoted beam. The circuit previously described is a series-fed Hartley oscillator; this one is a modified shunt-fed Hartley oscillator.

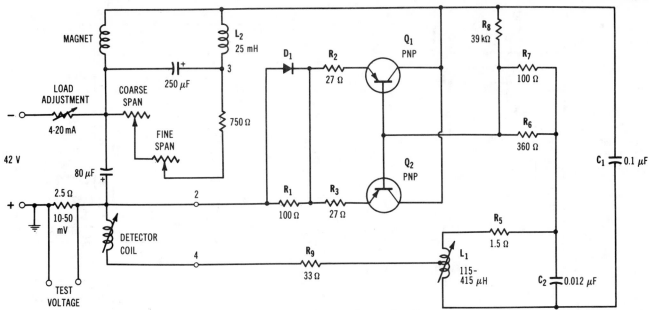

Fig. 10-3. Transmitter oscillator circuit.

Circuit operation will be discussed referring to the simplified schematic in Fig. 10-4. Coil L_2 represents a parallel combination of L_2 and the magnet unit from Fig. 10-3. Resistance R_E is a combination of the resistors in the emitter circuit, while resistance R_B is a combination of the resistors in the base circuit. When power is applied, collector current I_C will conduct through L_2, transistor Q, and the parallel combination of R_E and D_1. For this direction of current, D_1 has a very low resistance. The emitter is at practically ground potential. Resistor R_B develops a potential at the base that provides the proper forward bias. The variable inductor L_1 determines the frequency of the oscillator. The bandpass of the LC circuit is somewhat broadened by R_5 (see complete schematic in Fig. 10-3). This enables the circuit to oscillate over a broader range of frequencies than would otherwise be possible. The only major difference between the series-fed and the shunt-fed Hartley is the reference point or ground. With the series-fed Hartley (Fig. 10-2), the bottom of L_1 was grounded. This allowed the collector-emitter current to conduct through L_1. In the shunt-fed circuit, no collector-emitter current conducts through L_1 since the emitter is grounded.

As mentioned before, when power is applied, collector current conducts through D_1, Q, and L_2. A voltage drop is developed across L_2 with the polarity indicated. This voltage causes current to conduct through the detector coil and the lower portion of L_1. The current through this portion of L_1 then induces a voltage across L_1 that increases the forward bias on the transistor so that the collector current increases. This continues until the transistor saturates, the feedback cycle reverses, and oscillations are maintained in the L_1–C_2 tank

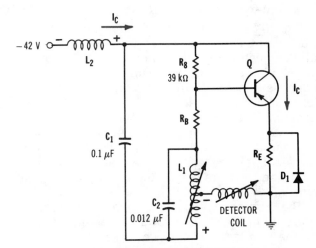

Fig. 10-4. Simplified transmitter oscillator circuit (shunt-fed Hartley).

circuit as described previously for the circuit of Fig. 10-2.

TRANSDUCERS

The purpose of a transducer is to change an input signal of one type into an output signal of a different type that is proportional to the original input. The type of output to be produced depends on the function we wish to perform. Four types of transducers are discussed in this section. In all but one case described here, the output is current, whereas the input is some kind of pressure signal. By transforming the input pressure to a 4–20-milliampere current, efficient long-range transmission of the signal can be accomplished to a standard recorder, indicator, or controller. The fourth transducer has a current input with a pressure output.

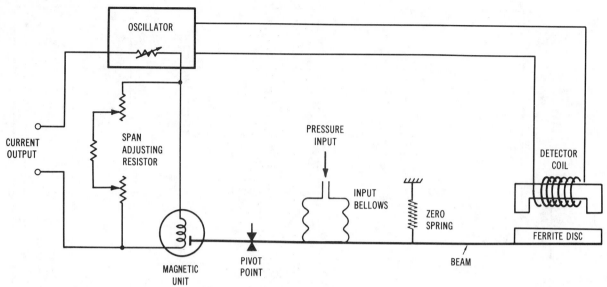

Fig. 10-5. Pressure-to-current transducer.

Pressure-to-Current Transducer

Fig. 10-5 illustrates the P/I (pressure-to-current) transducer. A balanced pivoted beam is the heart of this instrument. With no pressure input, the zero spring positions the beam. The position of the beam is sensed by the detector coil, the inductance of which changes as the beam changes position. The change in inductance is caused by the change in the air gap between the core of the coil and the ferrite disk. The magnet unit provides for feedback to rebalance the beam. The span-adjusting resistors control the amount of feedback current.

The operating sequence, with an increase in pressure in the input bellows, moves the right end of the pivoted beam down. This increases the air gap between the detector coil and the ferrite disk, causing a decrease in the inductance of the coil. The decrease in inductance changes the oscillator frequency, which results in an increase in current. This output current passes through the magnet unit and the span-adjusting resistors. The

amount of this current depends on the set value of the parallel span-adjusting resistors. The magnetic field produced by the feedback current through the magnet unit opposes the pressure input. The upward force of the left end of the beam produced by the magnet unit balances the beam at the new operating position. Any further change in pressure provides a corresponding change in output current.

Current-to-Pressure Transducer

Fig. 10-6 is a simplified schematic of the I/P (current-to-pressure) transducer. The sequence of operating steps follows: An increase in current produces a magnetic field across the magnet coil, the strength of which is determined by the position of the span-adjusting resistors. The reaction between the magnet coil and the permanent magnets within the magnet unit moves the beam upward. The movement of the beam increases the pressure on the nozzle, resulting in an increase in the air pressure going to the pilot valve. The output pressure and the pressure feedback also

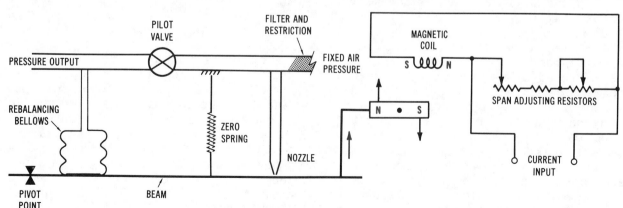

Fig. 10-6. Current-to-pressure transducer.

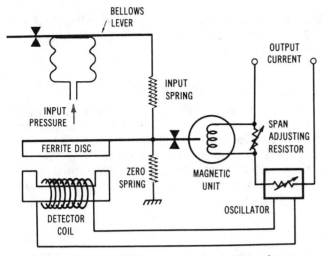

Fig. 10-7. Processed pressure-to-current transducer.

increase. The increase in pressure in the rebalancing bellows sets up a new operating balance point for the pivoted beam. The zeroing spring sets the balance point for a minimum current input. Any further change in current produces a corresponding change in pressure output.

Processed Pressure-to-Current Transducer

The PP/I (processed pressure-to-current) transducer is shown schematically in Fig. 10-7. This transducer is much like the P/I transducer. The primary differences arise from the fact that the processed pressures are relatively high. These higher pressures must be fed into the system through a high-pressure bellows or a Bourdon tube. Which of these devices is used depends on the range of pressures to be measured. The bellows positions a bellows lever that is connected to the pivoted beam through an input spring. Movement of the beam occurs when an unbalance results between the input spring and the zero spring. The repositioning of the beam causes a change

of inductance in the oscillator circuit. The increase results in a direct-current increase through the oscillator. A portion of this output current provides feedback to the beam through the magnet unit. The new balanced position of the pivoted beam is the operating point for this particular pressure. Any further change in pressure results in a beam unbalance. The same sequence of events is followed as the system adjusts to a new operating point.

Differential Pressure-to-Current Transducer

Fig. 10-8 illustrates the action that occurs in the differential P/I transducer. The differential pressure between the high- and low-pressure bellows is sensed by the liquid-filled torque tube. A change in differential pressure rotates the takeoff arm. This motion is transferred to the cross-spring beam through the input spring to the pivoted beam. The actions of the beam, oscillator, and magnet unit are exactly the same as for those discussed previously.

SUMMARY

An important part of a transmitter is the oscillator. Essentially, an electronic oscillator changes dc into alternating current. Through this part of the transmitter a small value dc input signal can be sent or transmitted to a distant location where it may achieve control or produce a signal indication.

A series-fed Hartley oscillator contains a tapped coil as an identifying component. This coil and a capacitor are used to form the LC frequency-determining components. In addition to this, there must be an amplifying device, dc power source, and a positive feedback loop.

The operation of a Hartley oscillator is based on the conduction of an amplifying device such

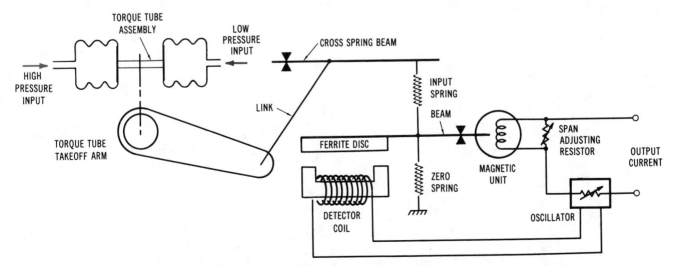

Fig. 10-8. Differential pressure-to-current transducer.

as a transistor. When dc power is applied, the transistor starts to conduct because of its forward-biased emitter-base junction. This current causes positive feedback from collector to base which charges a capacitor in the LC circuit and ultimately causes the transistor to saturate. When saturation occurs, feedback stops, the capacitor discharges, and the emitter-base junction becomes reverse biased. The discharging capacitor also causes an oscillatory current in the LC circuit. Ultimately the reverse-bias voltage is depleted and the transistor begins to conduct and repeat its operational cycle. The oscillatory frequency of the

LC circuit continues because of the flywheel effect. The frequency of the oscillator is based on the formula

$$f_o = \frac{1}{2\pi\sqrt{LC}}$$

The input to a transmitter is developed by a transducer. Essentially a transducer changes energy of one form into a different form. Pressure-to-current (P/I), current-to-pressure (I/P), processed pressure-to-current (PP/I), and differential pressure-to-current are some examples of instrumentation transducers.

CHAPTER 11

Recorders and Indicators

INTRODUCTION

In this chapter we will investigate recorders and indicators that have a wide range of applications in the industrial process control instrumentation field. Numerous companies make recorders and indicators that will drive a pen or produce a reading on a scale. Fig. 11-1 shows a representative indicator of the hand-deflection type. The

Courtesy Leeds & Northrup Co.

Fig. 11-1. A hand-deflection indicator.

scale is in millivolts, and switches on the lower panel are control functions actuated by the controller at various scale positions.

The first part of this chapter discusses recorder power supplies. This part of the recorder is primarily responsible for developing all of the ac and dc voltages utilized by the recorder. Secondly, we will discuss the pen servomechanism. This circuit controls the pen or stylus as it makes a continuous record on a chart. The third part is devoted to the error servo. This circuit provides an error signal for either an automatic or manual controlling device. The discussion will next be concerned with the switching arrangement. Provisions in this instrument have been made so that a changeover from automatic to manual operation can be accomplished smoothly. The final part of the chapter describes a method of instrument indication that has no moving parts. Indications are made on an electronic bar graph that is viewed from the front of the indicator.

RECORDER/INDICATOR POWER SUPPLY

Fig. 11-2 is the schematic diagram of the recorder power supply. The power transformer has four secondary windings, providing six dc voltage sources and two ac voltage sources for the recorder. The chart drive motor is connected in parallel with the power transformer primary. The primary voltage is 117 volts ac, 60 Hz. The first transformer secondary provides an ac voltage to silicon diodes D_{301} and D_{302}. The full-wave rectifier output of −42 volts dc is filtered by C_{301}. The second winding of the transformer provides a positive 50 volts dc by using diode D_{305} and filter C_{302A}. A center-tapped portion of this same winding provides a second −42-volt dc power source. This voltage is rectified by D_{303} and D_{304} and filtered by C_{307}. The −42-volt dc sources of the two full-wave rectifiers provide voltage for the pen servo and input circuit if the transmitter is not self-powered.

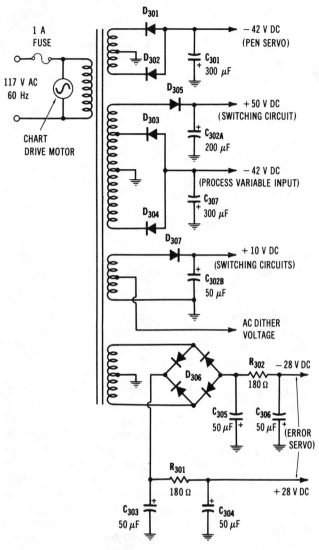

Fig. 11-2. Power supply for recorder section of Honeywell ElectriK Tel-O-Set.

The +50-volt dc half-wave power source is used in the switching circuits.

The third winding of the power transformer provides +10 volts dc to the switching circuits. Diode D_{307} is the rectifier for the power source, while capacitor C_{302B} is the filter. This is also a half-wave rectifier. A tap on this winding provides an ac dither voltage to the pen torque motor. The fourth winding is center-tapped and provides voltage to a full-wave bridge rectifier, D_{306}. The full-wave bridge rectifier is connected so as to produce both negative and positive outputs. Both 28-volt outputs are filtered by pi-section filters made up of a 180-ohm resistor and two 50-microfarad capacitors.

PEN SERVO

Fig. 11-3 is a functional block diagram of the pen servo. It is composed of a magnet unit, piv-

oted beam, detector coil, oscillator/amplifier, and torque motor. The action of the pivoted beam, detector coil, magnet unit, and oscillator is much the same as described in the chapter on transmitters.

The 4–20-milliampere output signal of the transmitter unit passes through the PV (process variable) coil. A magnetic force, which is due to the current through the PV coil and its reaction to the permanent magnets on the beam in the magnet unit, moves the pivoted beam. A change in beam position changes the inductance of the detector coil. With an increasing current input, the inductance of the coil increases. This is a result of the movement of the beam toward the detector coil. With a decrease in current, the detector coil inductance decreases. As you will remember, this inductance is in the oscillator feedback path. The change in inductance will change the oscillator amplitude and, hence, the collector current.

Fig. 11-4 is a schematic diagram of the oscillator/amplifier. You will notice that the circuit operation of the oscillator is identical to those used in the transmitters. The primary difference is that only one transistor is used in this oscillator circuit. The amplifier uses a single-stage power transistor. The input to this amplifier is developed across R_{473} by the oscillator collector current. This voltage is directly coupled to the base of the amplifier transistor, Q_{462}. The base voltage controls the collector current through Q_{462}. This current passes through the field winding of the torque motor, which reacts with the permanent magnet armature to position an indicator or recorder pen. A 5- to 10-volt ac dither voltage is constantly applied to this field winding to improve its dynamic response.

Referring again to Fig. 11-3, you will notice that feedback is obtained mechanically through a feedback spring connected between the recorder pen and the pivoted beam. Two kinds of damping are also used. Mechanical damping, also used in the transducers, is accomplished by an oil-filled dashpot attached to the beam. Electrical damping is accomplished by the use of a damping coil wound on the same core as the PV coil. This coil is connected in series with R_{471} and C_{464}, between the base and collector of Q_{462}. Any change in oscillator output is fed through this capacitor and, hence, through the coil. The magnetic field so produced counteracts the action of the PV coil, resulting in a damping or slowing effect on rapid beam movement which would cause vibrating or hunting in the circuit.

ERROR SERVO

Fig. 11-5 is a functional block diagram of the error servo. This circuit is made up of a magnet

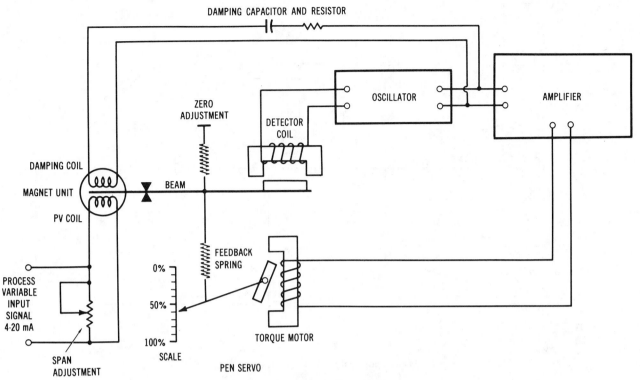

Fig. 11-3. Functional block diagram of pen servo.

unit, pivoted beam, detector coil, oscillator, set-point index, and set-point spring. The PV coil, which is a part of the magnet unit, is connected in series with the pen servo PV coil. Any transmitter input current to the pen servo also passes through the error servo PV coil. Such a current produces a magnetic field that moves the pivoted beam. Beam movement is detected by a change in

inductance of the detector coil. This change of inductance has the same effect on the oscillator as previously described.

A +28 volts dc and a −28 volts dc provide power for the oscillator circuit. These voltages are connected to the circuit in such a way that the oscillator forms one leg of a bridge circuit. The bridge is balanced when the oscillator acts as a

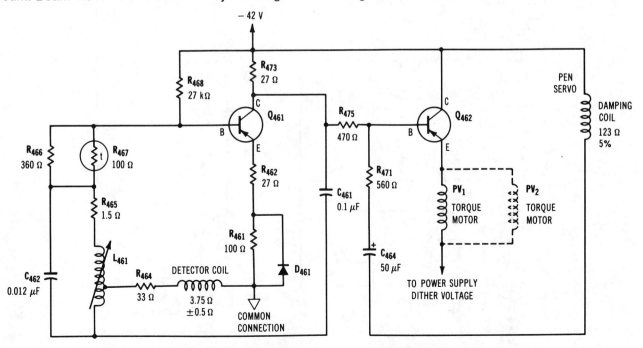

Fig. 11-4. Schematic diagram of the oscillator/amplifier.

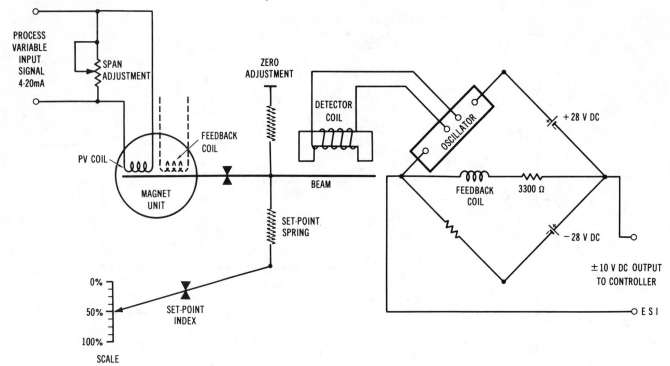

Fig. 11-5. Functional block diagram of error servo.

2200-ohm resistor. This occurs only when the PV current corresponds to the set-point index. When the PV current does correspond to the set point, there is no difference of potential across the bridge. The output is then zero. The feedback coil, which is wound on the same core as the PV coil, is also connected across the bridge. When the bridge is balanced, no feedback current is present.

A change in PV current produces an oscillator change. The change in oscillator current unbalances the bridge, which causes the voltage at ESI to change. The range of voltages across the bridge is ±10 volts dc. This difference of potential is the input to the controller. Controller action due to this input positions the final control device so that the PV coil current again corresponds to the set-

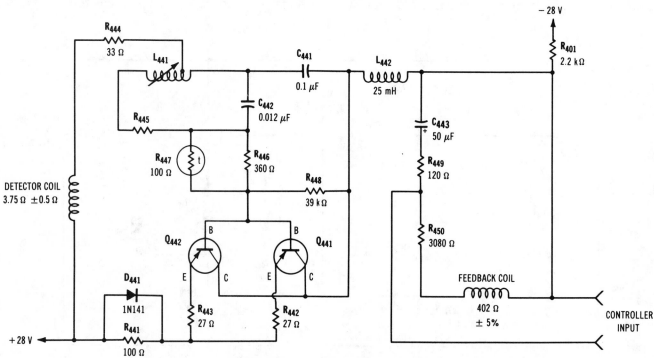

Fig. 11-6. Schematic diagram of error servo.

point index. A difference of potential across the bridge also causes current to conduct through the feedback coil. The magnetic field so produced repositions the beam.

Fig. 11-6 is a schematic diagram of the error servo.

AUTO/MANUAL OPERATION

Fig. 11-7 is a simplified schematic diagram of the auto/manual switching circuits. When the auto/manual switch is in the auto position, relay coil K_1 is energized, which causes the contacts, K_{1B}, to connect the output of the controller directly across the load that represents the final control device. Controller output, therefore, controls the position of this device. Meter M_1 indicates the amount of current being delivered by the controller. When this current is 12 milliamperes, the voltage drops across R_{201} and R_{202} are the same. The difference of potential across the meter is, therefore, zero. Under these conditions, the meter reads midscale, or 50%. Any change in controller current causes a deviation from this midpoint.

In order to switch to the manual position, the manual control, R_{210}, must be balanced. This is done by first depressing the momentary valve-balance switch to the balance position. When this is done, the 10-volt dc power source is removed

from the meter circuit and the voltage drop across R_{202} is produced by the transistor collector current. The manual control varies the base voltage and, hence, the collector current of the transistor, Q_{201}. This control is adjusted so that this current is the same as the controller output current. The auto/manual switch can then be thrown to the manual position, resulting in a smooth transfer. In the manual position, relay K_1 is de-energized. The contacts of the relay place the load in the collector circuit of Q_{201}. The meter is also connected into the manual circuit. The load can now be controlled manually by positioning R_{210}. A meter indication corresponds with this setting.

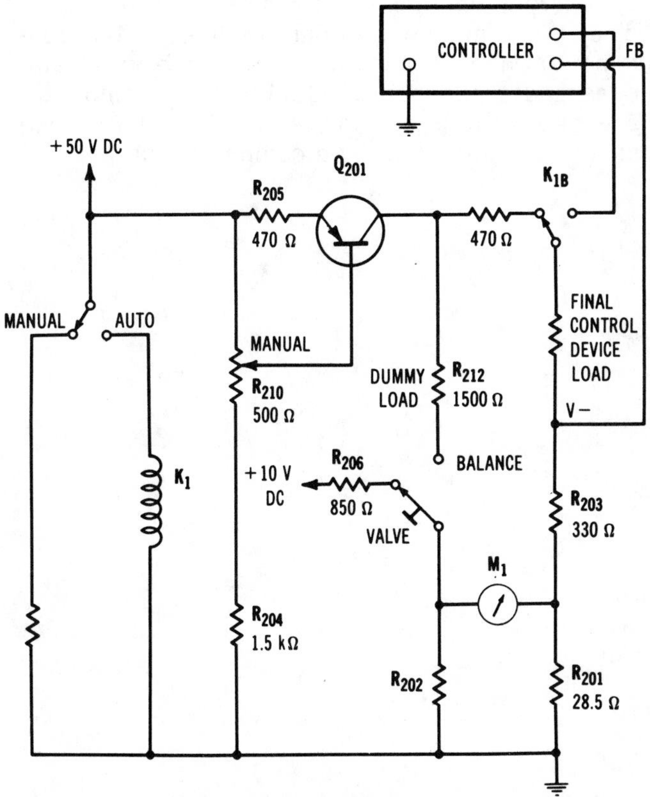

ELECTRONIC INDICATORS

The electronic indicator is an innovation that is unique when compared with the pen recorder or hand-deflection recorder. It has no moving parts and displays information on one or two bar-graph scales. The Moore Products Co. Model 375C Electronic Vertigage® unit, shown in Fig. 11-8, employs a digital principle of operation to provide an analog indication having an accuracy that was pre-

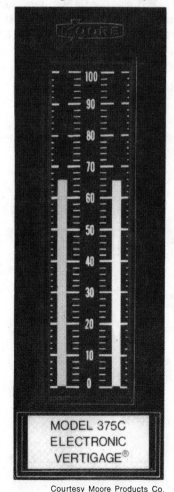

Fig. 11-8. An electronic indicator.

Fig. 11-7. Simplified schematic of auto/manual switching circuits.

viously associated with only servomotor-driven indicators.

Each input variable is displayed on a multisegment gas-discharge type of neon display tube. The display gives a bar-graph representation, the height of which is proportional to the input signal. Each display bar is approximately 0.1 inch (2.54 millimeters) wide and at full scale input it is expanded to 4 inches (101.6 millimeters) high. Numerous display scales are available to fit a wide range of indicator applications.

The bar-graph element shown has a right-hand and a left-hand bar each containing 200 cathodes that are 0.1 inch (2.54 millimeters) wide. Each bar has a single anode that is common to all of the 200 individual cathodes.

In operation, the cathodes are scanned from bottom to top with 60-Hz ac. On each scan from bottom to top, the individual segments are illuminated in sequential order until the anode is turned off by a comparator circuit. The comparator circuit is energized when the input signal coincides with a precise 1–5-volt dc sawtooth voltage signal. As a result of this action, the bar graph is illuminated to a height that is proportional to its input. Accuracy of this display method is ±0.5% of the 4-inch (101.6-millimeter) span, with a 0.2% linearity and a repeatability of 0.1%. With this type of indicator, mechanical maintenance problems are eliminated.

SUMMARY

A wide range of recorders and indicators are available today in the industrial process control field.

Typical power supplies produce dc voltages of different values that range from −28 volts to +50 volts dc. Half-wave, full-wave dual-diode, and full-wave bridge rectifiers are commonly used to develop the necessary supply voltages. RC and pi-section filtering are commonly employed in the different power supply sections.

A pen servo is composed of a magnet unit, pivoted beam, detector coil, oscillator/amplifier, and a torque motor. A signal passing through the PV (process variable) coil moves the pivoted beam. A change in beam position changes the inductance of the detector coil and ultimately the oscillator frequency.

The error servo employs a magnet unit, pivoted beam, detector coil, oscillator, set-point index, and set-point spring. Any input signal to the pen servo passes through the error servo coil. This current produces a magnetic field that moves the pivoted beam, which causes an inductance change and ultimately a frequency change of the oscillator. When the PV input corresponds with the set-point index, balance occurs. Changes in the PV input alter the oscillator frequency, which unbalances the bridge.

When a controller is in the automatic position, relay contacts place the output of the controller directly across the load. Controller changes influence the load directly when it is in this position of operation.

Electronic indicators have no moving parts and display information on a bar graph. Each bar contains 200 neon display lamps with separate cathodes and a common anode. When the lamps are scanned with ac they will be illuminated until one segment is turned off by a comparator circuit.

CHAPTER 12

Electronic Controllers

INTRODUCTION

An electronic controller is the instrument that is primarily responsible for industrial automation (the automatic control of process variables). Applications range from simple on/off control operations to completely automated systems that respond to signals that are initiated by direct access to a digital control computer. Control may apply to only one variable such as temperature, pressure, electrical conductivity, fluid flow, etc., or it may respond to literally hundreds of process variables simultaneously.

In this chapter we will look at the electronic circuitry of some representative industrial controllers. Electronic controllers have been used in industry for a number of years and have gone through several transitions. When process controllers were first introduced in the early 1940s, they were vacuum-tube devices mounted in large metal cabinets. With the development of solid-state devices in the late 1950s, the outward appearance of the controller had some significant changes. The

transition to all solid-state controllers, however, has been rather slow. Initially most companies were reluctant to give up their popular selling vacuum-tube controllers. There was then a period when controllers employed both vacuum tubes and solid-state devices. These "hybrid" controllers and many of the vacuum-tube devices are still in operation today. Some manufacturers still have vacuum-tube controllers available because the demand for them continues to be surprisingly good.

All major controller manufacturers today produce a wide variety of solid-state instruments. These units, in general, are small in size and

Courtesy Moore Products Co.

Fig. 12-2. An indicating controller partially removed from its housing showing plug-in modules.

Courtesy Leeds & Northrup Co.

Fig. 12-1. A typical solid-state controller.

usually employ hundreds of discrete components. Fig. 12-1 shows a typical solid-state controller of this type. It has unusually precise control capabilities with exceptional stability.

The advent of solid-state controllers has brought about some innovative design features that have had a decided impact on controller maintenance. Components, for example, are mounted on removable printed-circuit cards or boards for easy replacement. Fig. 12-2 shows an example of an indicating controller that has been partially removed from its metal housing. The printed-circuit modules of this controller can be easily removed by pulling the wire rings near the center of the controller. This controller has a great deal of versatility through this type of construction. Fig. 12-3 shows a functional diagram of the potential location of alternate modules that can be utilized in this unit.

The next trend in controller technology found large numbers of discrete solid-state components replaced by integrated circuits. These controllers are somewhat smaller than their discrete component solid-state counterparts. Maintenance, in this case, is based on faulty IC determination and PC board replacement.

Microprocessors are now finding their way into the process controller field. A microprocessor has a large number of discrete ICs on a single chip. In many situations the entire controller may be built on a single chip. This will obviously change the controller size again, and in many cases eliminate most maintenance problems.

With the wide range of diversity that exists today in controller technology, it is difficult to single out a particular controller that is representative of the field. In this regard, we will first discuss some common solid-state controller circuitry, then show some typical IC applications using op amps. Through this approach you will be able to pick out the information that is particularly applicable to your controller needs.

CONTROLLER FUNCTIONS

General Information

Fig. 12-4 is a block diagram of a typical controller. Three modes of operation are incorporated in this instrument: (1) proportional action, (2) reset action, and (3) rate action. They will be discussed in connection with the input circuits of the controller.

Sections will be devoted to the rate action, differential, output, and feedback amplifiers. The controller power supply will not be covered sep-

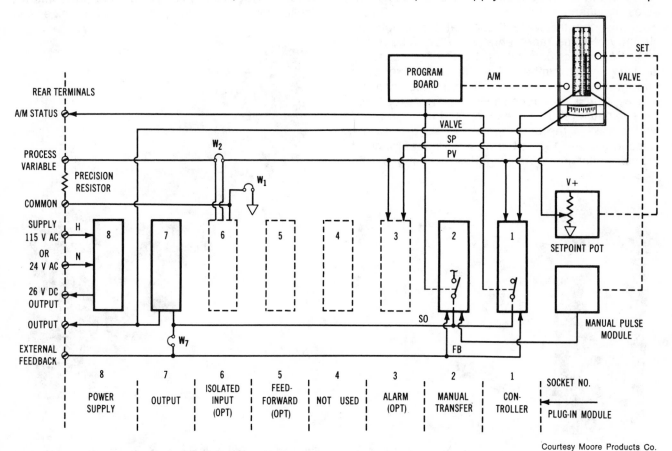

Courtesy Moore Products Co.

Fig. 12-3. Functional diagram of alternate module locations for the controller of Fig. 12-2.

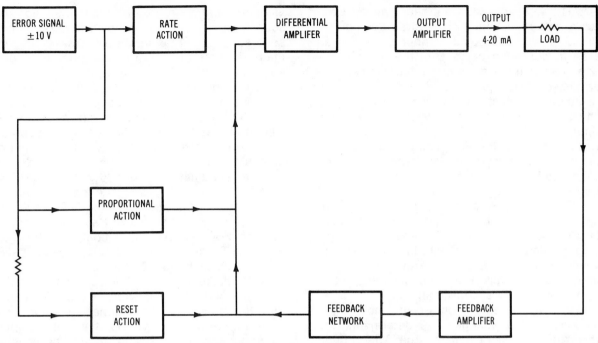

Fig. 12-4. Block diagram of a typical controller.

arately because it is so similar to power supplies previously discussed.

Controller Input Circuits

Fig. 12-5 is a simplified schematic diagram of the controller input circuits. Each of the three previously mentioned modes of controller operation are described here.

1. **Proportional Action.** This action determines the ratio between the controller output signal and the input signal. If the proportional-band control is set at 100%, the controller output will result in a change that is directly proportional to the error signal. For insensitive transmitters, the proportional-band control will be set for less than 100%. The error signal will, therefore, produce an output that is proportionally greater. This is called narrow-band control. A narrow range in error signal can produce a full range of controller outputs. For settings of the proportional-band control of greater than 100%, the error signal range will be greater than the controller output. This is called wide-band operation.

2. **Reset Action.** This action is constantly driving the final control mechanisms to zero-out any error signal. Any time the controller position differs from the set point, reset action moves the controlling device in such a direction as to agree with the set point. The amount of action depends on the amount and the length of time of the deviation.

3. **Rate (Derivative) Action.** This action determines the rate of controller action. Its effect

on controller output is twofold: If the controller output rate were dependent on proportional-band control only, it is possible for the error signal to become so great that the controller could not possibly zero itself. On the other hand, if the rate action were too fast, the controller would oscillate or hunt. The error signal input to the controller is differentiated so that its rate of change can be detected and the proper rate action provided.

Fig. 12-5 shows that the error signal is applied to a voltage divider made up of R_{25} and R_{26}. This signal is a dc voltage with a magnitude of 10 volts

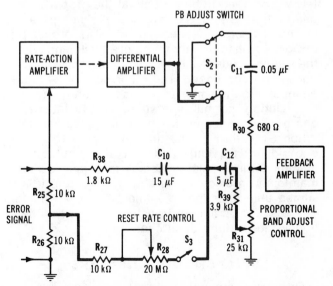

Fig. 12-5. Simplified schematic diagram of controller input circuits.

or less. The plug into the controller can be positioned in either the direct or reverse position. In the direct position, a positive input produces an increase in controller current. In the reverse position, a positive input produces a decrease in controller current. This input is fed directly to the rate-action amplifier, which will be discussed later. Its output is one of the inputs to the differential amplifier.

The first input to the differential amplifier is a signal that combines reset and proportional action. One-half of the error signal is developed across R_{26}. This voltage is fed to the reset control circuit. The reset control circuit is made up primarily of R_{28} and C_{12}. R_{27}, R_{39}, and a portion of R_{31} are also in the circuit. Since R_{28}, calibrated in repeats per minute, is so much larger than the others combined, it largely determines the time constant of the circuit. This time constant determines how often the proportional response is repeated. In other words, C_{12} charges to the change in the error-signal voltage; R_{28} determines the length of time it takes C_{12} to charge. The voltage across C_{12} is the second input to the differential amplifier. The amplifier is stabilized by a feedback voltage that is equal and opposite to the input. How this signal is produced will be discussed in the section covering the feedback amplifier.

Proportional action is determined by the percentage of feedback-amplifier output that is applied to the differential-amplifier input. The percentage of feedback is determined by the setting of the proportional-band control, R_{31}. The voltage at the wiper arm of R_{31} determines proportional action as well as a reference voltage for C_{12}. The reset action voltage, the charge across C_{12}, adds to this voltage. The sum of these two voltages makes up the input to the differential amplifier. A few words need to be said about the proportional-band adjust switch, S_2. This switch is incorporated in the proportional-band adjust (R_{31}) procedure by being a push-and-turn type of adjustment. When the switch is pushed, the normal differential-amplifier input is grounded. The other pole of S_2 feeds an input into the amplifier that is the charge across C_{11}. During normal operation, C_{11} maintains a charge that is equal to the voltage across R_{31}. When S_2 is thrown, the ground is removed from one side of C_{11} and applied to the input of the amplifier. Since this voltage is the same as that maintained previously, no change in output is assumed until after the adjustment has been made. The result is a smooth transfer from one proportional band setting to another.

SOLID-STATE CONTROLLERS

Discrete component solid-state controllers have obviously not been around as long as their older vacuum-tube counterparts. Nearly all controllers sold in recent years are, however, predominantly solid state. Fig. 12-6 shows a representative schematic diagram of a discrete component solid-state controller.

As can be seen in Fig. 12-6, the power supply provides two dc outputs. They are +36 volts and +46 volts, respectively. Each power supply uses silicon diodes to provide full-wave rectification.

Filtering of the power supply voltage is not quite as obvious in this circuit as is generally displayed in other schematics. The +36-volt source, for example, is filtered by a pi-section filter near the center of the diagram. (See printed-circuit points 32, 33, and 10.) The +46-volt supply, by comparison, employs an LCR filter composed of L_2, C_{12}, and R_{31}. These components are located between printed-circuit points 27 and 26.

In the discussion that follows, we will single out specific solid-state controller circuits for explanation. In some cases operational amplifiers will be used in circuits where this type of component finds acceptance today. Through this approach you will see how the IC simplifies the circuit and the explanation of its operation.

The controller input is the error voltage developed by the recorder. This input signal is negative or positive, depending on whether the process variable signal is above or below the set point. When the process variable signal is the same as the set point, the input to the controller is zero. The operational amplifier circuit shown in Fig. 12-7 is the basis for amplifier action. It consists of an input network and a feedback network in which the current is equalized by an amplifier. If the input current is constant, C_{in} charges to that voltage. The feedback current, produced by this input voltage, charges the feedback capacitor, C_f. Once the two capacitors are equally charged, no current will conduct. The slight excess of input current necessary to sustain amplifier output is insignificant.

An input signal causes an input current to conduct, charging C_{in}. This amplifier input immediately produces an output, or feedback current. Feedback current charges C_f to a value that equals the charge on C_{in}. Since the reaction of the amplifier is practically instantaneous, the feedback current is constantly tracking or following the input signal. As these two equal currents vary, adding to and subtracting from the charge of the capacitors, these charges always remain the same. As a result of this action, the junction of the two capacitors (summing junction) is always essentially at a ground potential.

Proportional Action

The effective ratio of the value of the input capacitance and the feedback capacitance deter-

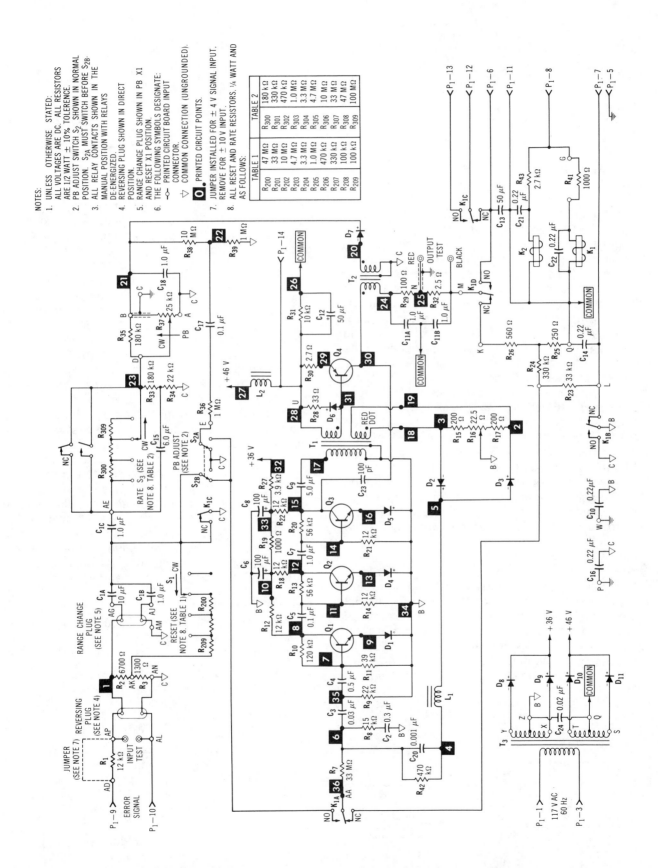

Fig. 12-6. Schematic diagram of a typical discrete component solid-state controller.

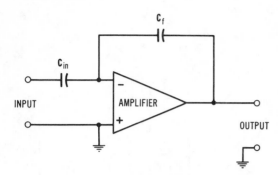

Fig. 12-7. Operational-amplifier equivalent of
transistorized controller.

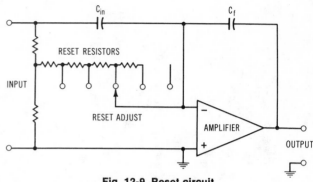

Fig. 12-9. Reset circuit.

mines the proportional band. The effective value depends not only on capacitor size, but on the voltage gain of the amplifier. If the effective ratio is 1:1, then the proportional band is 100%. Under these conditions, a full-scale input change produces a full-scale output change. A change in the ratio of the capacitances changes the proportional band. A range of proportional bands can be provided by 1–100% or 10–1000% by changing the ratio. In addition to a variation in range of the proportional band, a change within the range is provided by a potentiometer. The potentiometer is the proportional-band adjust control shown in Fig. 12-8. By adjusting this control, the amount of the output that is used in the feedback circuit can be varied. With a decreasing feedback signal, a larger output must be produced to fully charge C_f. An increased output for a given input is merely a narrowing of the proportional band. A full range of output signals can, therefore, be produced by small input signals. The range of input in percent that produces a full range of output signals is the proportional band.

Reset Action

A voltage divider is placed across the controller input and a variable resistor is placed in parallel with C_{in} for reset action (Fig. 12-9). The voltage divider and variable resistor provide a path for current whenever an error signal exists. This current provides a continuous amplifier input whenever the process variable signal differs from the

set point. Feedback current, therefore, continues to charge, or discharge, the feedback capacitor as long as there is an input. The amplifier then produces a continuous change in output as long as an input signal exists. This, by definition, is reset action, since the controlled device is seeking the null position. In discussing the proportional circuit, it was evident that there was current only when there was a change in the input signal. In the reset circuit, the only time that input current stops is when the error voltage is zero. By decreasing the value of the reset resistors, input current is increased for a given input signal. The output current must increase at a faster rate in order to produce a matching charge across the feedback capacitor. The result is a faster reset rate. Increasing the size of the reset resistors results in a slower reset rate. To produce a very low reset rate, the voltage divider has been incorporated in the input circuits. The voltage divider causes the reset resistors and the amplifier to see a smaller portion of the input signal, resulting in a smaller charging current. The input capacitor, however, still sees the full input voltage. The value of the reset resistor determines the number of times that proportional action is repeated per minute.

Rate Action

Rate (derivative) action is accomplished by the addition of a voltage divider, a variable resistor, and a capacitor in the feedback circuit (Fig. 12-10).

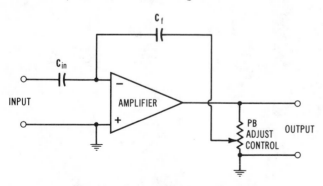

Fig. 12-8. Proportional-band circuit.

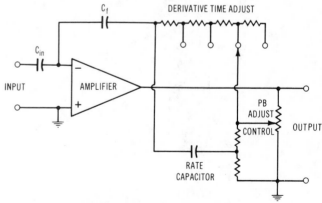

Fig. 12-10. Derivative (rate) circuit.

An input signal change produces an output change that is easily passed by the rate capacitor. Since this change is taken from across the voltage divider, the result is that the feedback capacitor sees only a portion of the output. The effect is the same as a decrease in the proportional-band control. That is, the instantaneous effect is a narrowing of the proportional band. This results in a greater output change for a given input until the rate capacitor charges. The time it takes the rate capacitor to charge depends on its time constant. This can be changed by the variable resistor, the rate time adjustment. As soon as the rate capacitor charges, the effect of the voltage divider disappears. The proportional band is then restored to its original value. Rate amplitude will be determined by the voltage divider. This, of course, determines the amount of the instantaneous change in signal fed through the rate capacitor.

Rate action has an anticipatory function that acts only when the output is changing. It then provides a braking or damping action. In general, rate action is used to reduce cycling and overshoot and to permit the use of narrower proportional bands.

Amplifier

The amplifier used in the transistorized controller can be divided into three basic sections: the impedance bridge and internal feedback loop, the three-stage oscillating amplifier, and the output stage. The schematic for the amplifier is shown in Fig. 12-11.

A Wheatstone or impedance bridge forms an important part of the feedback loop that makes oscillation within the amplifier possible. This bridge is made up of two fixed resistors, a potentiometer, and two diodes. Two legs of the bridge are resistive, while the two remaining legs are formed by the diodes. The diodes provide rectifying action and also function as variable resistances. The amplifier output is applied across the bridge. A portion of this voltage developed across the bridge is fed to the amplifier input. The amount of feedback is determined by the position of the bridge potentiometer. The polarity of output coupled to the feedback loop is determined by the feedback winding of the output transformer. The polarity is such as to provide positive feedback (regeneration) at the operating frequency. The amount of feedback is determined by the bridge as mentioned previously. The frequency of oscillation, about 20 kHz, is determined by the coil and capacitors located in the amplifier input circuit.

The input signal comes from the recorder. This voltage is fed to the base of Q_1 along with the feedback signal. If this error signal has the same polarity as the feedback signal, the output of the three-stage amplifier will increase. On the other hand, if the polarity is reversed, the output voltage across transformer T_1 will decrease. The amplitude of the signal across the primary of T_1

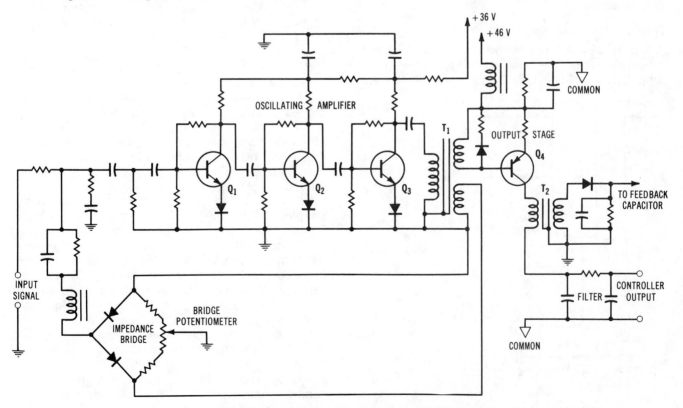

Fig. 12-11. Oscillating amplifier and output circuit.

depends on the error signal. The three-stage amplifier is composed of three npn transistors connected as common-emitter amplifiers. The three stages are RC-coupled, and a voltage divider determines the forward bias of each.

The base signal is taken from a secondary winding of T_2, which is associated with Q_4, a pnp transistor that is the output amplifier. Transistor collector current passes through the load for controller output. Since this current is a changing or pulsating dc, it must be filtered to give a dc current range of 4 to 20 milliamperes. Collector current also passes through the primary of transformer T_2. The secondary voltage, rectified and filtered, charges the feedback capacitor as discussed in the preliminary paragraphs of this section.

VACUUM-TUBE CONTROLLERS

Using the previous controller block diagram (Fig. 12-4) as a general reference, vacuum tubes can be used to achieve the same basic modes of controller operation as those of the transistor circuits. In general, these circuits are considered obsolete today. A large number of vacuum-tube controllers are, however, still being used in many applications. For those who have a need for vacuum-tube circuit information, continue on with the following circuit investigation. Feel free to pass over this material if there is no immediate need for vacuum-tube controller applications in your line of work.

Rate-Action Amplifier

Fig. 12-12 is a simplified schematic of a vacuum-tube rate-action amplifier. This is a two-stage dc amplifier. The first stage is a normal voltage am-

plifier and the second stage is a cathode follower. A dc plate voltage for both stages is provided by 100 volts ac, which is rectified by D_1 and filtered by C_1. A negative voltage is developed across R_3 by rectifier action and is maintained at the cathode of V_{1B} by the action of C_3 and D_2. This develops the proper plate voltage so that direct coupling can be used with the proper bias on V_{1A}. V_{1B} amplifies and inverts the input error signal. The error signal input is integrated by the action of R_{24} and C_2. By integration, we mean that the rate of change in error signal voltage is averaged out by the rate of charge (time constant) of the input circuit. If the position of C_2 and R_{24} were reversed, the input signal would be differentiated. In both cases, the amount of integration, or differentiation, is dependent on the size of the resistor or capacitor involved.

This changing signal is fed to the grid of V_{1A}. Static tube current maintains a constant positive voltage at the cathode of V_{1A}. C_4 assumes a charge equal to the voltage drop across R_4 (the cathode voltage). During static operation (no signal input) there is no current through R_5. The input to the differential amplifier is then zero volts during this period of time. When a change of voltage occurs at the grid of this cathode follower, the cathode voltage changes. Since the charge across C_4 does not change immediately (notice the long time constant of C_4 and R_5), the change appears across R_5. Notice that only the change, or rate of variation in voltage, appears as an output of the rate amplifier. R_5 and C_4 make up a differentiating circuit.

Differential Amplifier

Fig. 12-13 is a simplified schematic diagram of a vacuum-tube differential amplifier circuit. Two stages of amplification are provided for the rate-action signal. Two stages also amplify the combined proportional-band and reset-action signal. These two outputs are mixed by common cathode coupling; a single output is developed that is the difference of these two input signals.

The output of the rate-action amplifier is directly coupled to the grid of V_{2B}. Any change in tube current develops an output signal across plate-load resistor R_8. R_7 is a common-cathode resistor for both V_{2B} and V_{2A}. A change in tube current of V_{2B} changes the cathode voltage of V_{2A}. A cathode variation provides a signal input equal in amplitude, but opposite in polarity, to a grid signal. A plate-voltage change of V_{2A} results in a rate-action input. R_6 is a balance control that eliminates the need for perfectly balanced tubes. The output of V_{2B} is directly coupled to the grid of V_{3A}. The unbypassed cathode again feeds a signal to the common cathode of V_{3B}. This cathode signal develops an output across R_{15} that is fed to the

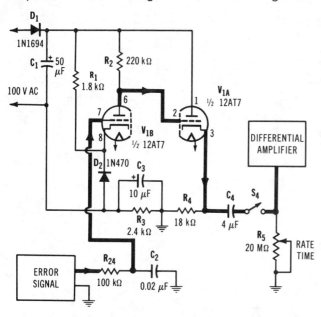

Fig. 12-12. Vacuum-tube rate-action amplifier.

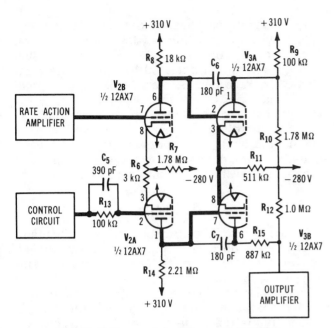

Fig. 12-13. Vacuum-tube differential amplifier.

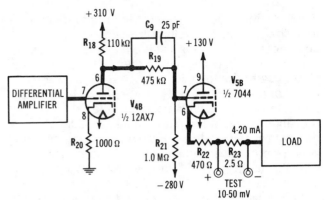

Fig. 12-14. Vacuum-tube output amplifier.

output amplifier, to which an input from the control circuit is fed directly through two stages of amplification. When two inputs occur simultaneously, the output of V_{3B} is a combination of the two inputs and is proportional to the difference in the two.

Output Amplifier

The vacuum-tube output amplifier (Fig. 12-14) consists of two stages of amplification involving tube V_{4B}, a voltage amplifier, and tube V_{5B}, a cathode follower. The input signal from the differential amplifier is fed to the grid of V_{4B}, which amplifies and inverts it. The output of V_{4B} is dc-coupled to the grid of the cathode-follower stage. Capacitor C_9, by shunting R_{19}, helps prevent parasitic oscillation. The cathode-follower stage takes the high-impedance voltage input from V_{4B} and changes it into a high-current, low-impedance output. The current output lies within the range of 4 to 20 milliamperes dc. A 2.5-ohm resistor (R_{23}) is inserted in series with the current output to provide a convenient test point to measure the output current in equivalent millivolts.

Feedback Amplifier

Fig. 12-15 is a simplified schematic diagram of a vacuum-tube feedback amplifier. The output-amplifier current passes through the load (0–1150 ohms) and a recorder resistor. This recorder resistor ensures a constant current output from the controller, regardless of load impedance. It also

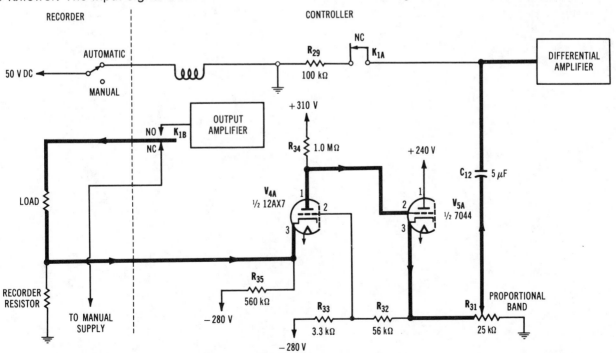

Fig. 12-15. Vacuum-tube feedback amplifier.

develops the input voltage for the feedback amplifier. The feedback amplifier involves two stages of amplification. The first stage (V_{4A}) is a normal voltage amplifier that is dc-coupled to the second stage, cathode follower V_{5A}. The voltage developed across the recorder resistor is fed to the cathode of V_{4A}. A cathode input affects tube current in exactly the same manner as if fed to the grid with an opposite polarity. A change in cathode voltage produces a plate voltage change in phase with the input signal. The plate-load resistor, R_{34}, is connected to the +310-volt plate supply, while the cathode resistor is connected to the −280-volt supply. By using these voltages, the plate voltage is maintained at about zero volts so that dc coupling can be used.

The output of V_{4A} is fed to the grid of V_{5A}. Any change of grid voltage produces an in-phase change in cathode voltage. Tube current develops this voltage across R_{31} (the proportional-band control) and the parallel voltage divider, R_{32} and R_{33}. The portion of the output voltage developed across R_{33} is fed back to the grid of V_{4A}. This voltage is out of phase with the input to V_{4A} and is, therefore, a negative feedback. The amount of negative feedback determines the gain of the feedback amplifier and stabilizes the dc amplifier. The amount of feedback at the input of the differential amplifier is determined by the setting of the proportional-band control. The less the feedback voltage, the greater the sensitivity of the entire system. This, of course, is the purpose of the proportional-band action.

As described previously, when the recorder is in the manual position, the load current is determined by the set position of the manual control. The controlled output is open-circuited. Manual load current is still detected by the voltage drop across the recorder resistor. This voltage is still fed to the input of the feedback amplifier. Feedback amplifier output maintains a charge across C_{12}. C_{12}, therefore, remembers the load current (the charge of C_{12} is proportional to load current) so that smooth, continuous operation can be maintained when switching to automatic operation.

SUMMARY

Today, electronic controllers are primarily made of solid-state devices employing discrete transistors or integrated circuits. Older controllers employed vacuum tubes.

The operational modes of a controller include proportional action, reset action, and rate action.

Proportional action determines the ratio between the controller output signal and the input signal. If a controller of this type is set at 100%, changes in output are directly proportional to the error signal. Insensitive controllers produce an output that is proportionally greater than 100%.

Reset action is used to constantly drive the final control element to zero-out any error signal that occurs. Any difference between set point and process variable input causes the controller to move toward the set point.

Rate-action control occurs when the error signal input is differentiated so that its rate change can be detected and some proper action provided.

CHAPTER 13

Converters

INTRODUCTION

The mV/I (millivolt-to-current) converter (Fig. 13-1) and the mV/P (millivolt-to-pressure) converter will be discussed in this chapter. Both circuits produce an electrical current output with a range of 4–20 milliamperes. The mV/P converter, however, incorporates a current-to-pressure transducer so that a final pressure output results. Because the electrical systems of the converters are identical, they will be discussed simultaneously. The I/P transducer will be discussed in Chapter 17.

The converter will be described in three sections. The first section involves the measuring circuit and its associated power supply, which incorporates a zener diode that provides a regulated dc suppression voltage. The measuring circuit can handle a standard millivolt input or a thermocouple input. When the thermocouple version is used, cold-junction compensation is provided. From the measuring circuit the signal passes through a filter and a vibrating-reed converter. In this part of the circuit, stray ac pickup is filtered out, the dc input is changed to a 60-Hz voltage, and negative feedback from the amplifier is fed into the circuit to produce an error signal.

The third section involves an amplifier and demodulator circuit. The amplifier in the converter has a very high gain. Its input is an error signal that is only about 0.5% of the measured signal. The advantages of using the high-gain amplifier with negative feedback have been discussed previously. You will remember that it minimizes variables within the amplifier and loading of the measuring circuits. The converter output is a current of 4–20 milliamperes that is proportional to the millivolt input.

MEASURING CIRCUIT AND POWER SUPPLY

Fig. 13-2 is a simplified schematic of the input or measuring circuit. You will notice from the diagram that two voltages are added algebraically to the input millivolt signal. The first of these voltages is added to the input by a bridge circuit in series with the negative input terminal. The reference or suppression voltage for this bridge is provided by a zener-diode-regulated power supply that makes the use of a standard cell or constant standardization unnecessary. The values of R_1, R_2, R_3, and R_6 vary with the range of the input. P_1 is a zero-adjust potentiometer. With a millivolt in-

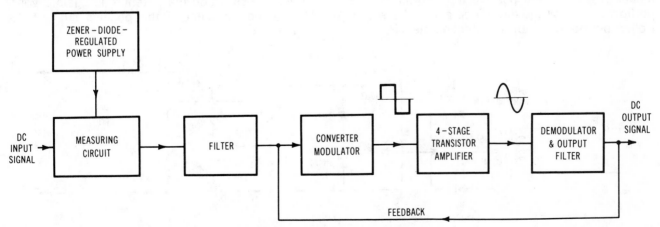

Fig. 13-1. Block diagram of mV/I converter.

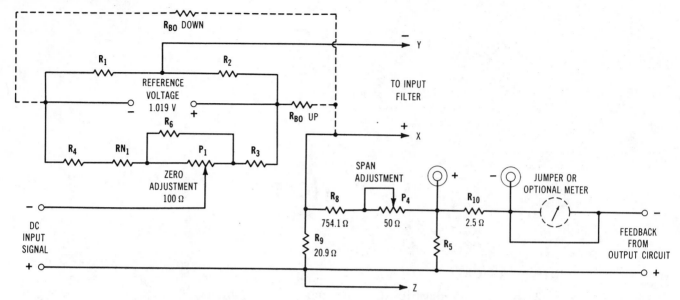

Fig. 13-2. Simplified schematic diagram of measuring circuit.

put, the value of RN_1 is zero. Under these conditions, the bridge adds a signal to the millivolt input so that the amplifier input range is within the capabilities of the instrument.

With a thermocouple input, RN_1 provides for cold-junction compensation. This resistor is wound around the cold-junction compensating block. A voltage drop across this resistor subtracts a voltage corresponding to the ambient temperature. R_{BO} provides protection in the case of a thermocouple failure. Fig. 13-2 shows the position of this resistor for either upscale or downscale deflection. The bridge circuit adds to the negative input potential a voltage that compensates for ambient temperature and differences in range.

Feedback voltage is added algebraically to the potential at the positive input terminal. Output current develops the negative-feedback voltage across R_5. This voltage is also developed across the parallel voltage divider made up of P_4, R_8, and R_9. The voltage across R_9 subtracts from the potential present at the positive input terminal. The portion of the voltage developed across R_9 is controlled by the span-adjust potentiometer, P_4.

The zener-diode-regulated power supply is shown in Fig. 13-3. This supply provides the 1.019 volts necessary for the operation of the measuring circuits. The power supply input is taken from a 90-volt center-tapped winding of the power transformer. Full-wave rectification is accomplished by diodes D_6 and D_7. The rectifier output is filtered by C_9. This dc output is applied to a voltage divider made up of R_{19} and two zener diodes. A zener diode always maintains a constant voltage across it. The zeners used in this power supply produce a regulated 7 volts. Two of these in series produce 14 volts across the voltage divider made up of R_{12} and R_{13}. The potential between R_{12} and R_{13} is the negative input to the bridge. R_{cu} provides for temperature compensation. P_2 adjusts the standardizing potential. This adjustment needs to be checked only biannually. The positive potential for the bridge circuit comes from a second stage of regulation. The 14-volt output of the power supply is developed across R_{11} and another zener diode, across which the potential is again 7 volts. This is the potential at the positive side of the bridge circuit.

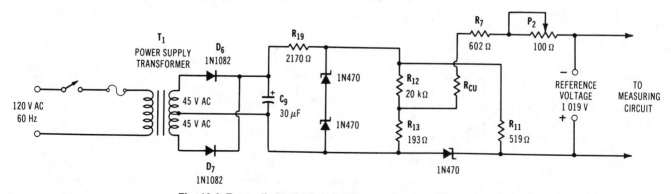

Fig. 13-3. Zener-diode-regulated reference voltage power supply.

A zener voltage regulator is produced by connecting a reverse bias across a pn junction. If this voltage is high enough, a breakdown of the diodes occurs and current conducts. The voltage drop so produced will be constant and is called the "zener voltage." The value of this voltage is determined by how the pn junction is formed. The zener voltage can be raised by diffusing the pn junction; that is, the higher the zener voltage, the wider the area in which there is a gradual transition from a p-type to an n-type crystal.

FILTER AND INPUT CONVERTER

Fig. 13-4 illustrates both the filter and chopper circuits. The filter is used primarily to eliminate stray ac pickup that may have occurred in the measuring circuit. It also stabilizes the loop. The unstable conditions that occur in the chopper are due to the changes of impedance when the vibrating reed makes and breaks contact. The chopper itself operates the same as those used in the instruments previously discussed. The reed, polarized by a permanent magnet, vibrates in synchronization with the 60-Hz current in the energizing coil. As a result, the dc input will produce a current first in one half of the input transformer primary, and then in the other half. The primary current produces an alternating voltage across the transformer secondary. This voltage is the amplifier input and closely resembles a 60-Hz square wave. Its amplitude depends on the dc input-error voltage.

AMPLIFIER AND OUTPUT CIRCUITS

Fig. 13-5 is a schematic diagram of the amplifier and output (demodulator) circuits. Four stages of voltage amplification are produced in this circuit. A very high gain is produced in this amplifier. All four stages of amplification use common-emitter, pnp transistors. The first and second stages are dc coupled, as are the third and fourth stages. The second and third stages are capacitance-coupled. Transistor Q_3 incorporates a gain control in its base circuit. The first three transistors develop their outputs across a resistive load in the collector circuit. Q_4 uses the primary of an output transformer as its collector load; collector current through this primary produces a voltage across the secondary. This voltage is the input to the output or demodulator stage.

The amplifier power supply produces about −32 volts dc. The power supply input comes from a 200-volt ac center-tapped secondary of the power transformer. Full-wave rectification is provided by diodes D_1 and D_2; filtering is accomplished by capacitors C_{7C}, C_{7B}, and C_{7A}, and resistors R_{41}, R_{17}, and R_{27}.

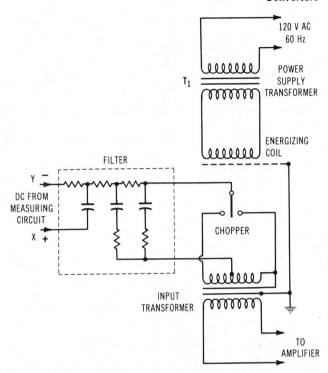

Fig. 13-4. Filter and chopper circuits.

A push-pull output stage is composed of transistors Q_5 and Q_6. The amplifier output is applied to the bases of these transistors by the center-tapped secondary of the output transformer. The collector voltage of these transistors is provided by an 80-volt center-tapped power transformer secondary. Half-wave rectification of this voltage is provided by D_8 and D_{11}. As long as the base voltage and collector voltage are in phase, the collector will conduct current. The amount of this current, of course, depends on the amplitude of the base voltage. Transistor Q_5 will conduct during one half-cycle, and Q_6 will conduct during the other half-cycle. This pulsating current is filtered by capacitor C_5. As a result, the dc output current is the average collector current of Q_5 and Q_6.

MILLIVOLT-TO-PRESSURE CONVERTERS

Millivolt-to-pressure (mV/P) converters serve as a link between an electrically energized controller and a pressure-actuated system. Fig. 13-6 shows a representative electric-to-pneumatic type of converter. This converter, or transducer, is essentially a motor-operated pressure regulator which converts varying dc signals into 3 to 15 lb/in² (20.7 to 103.4 kPa) air-pressure signals. Design features are specifically aimed at eliminating abrupt process upsets. This includes such things as self-locking valve position settings with power failure, and smooth valve adjustment without continuous balance adjustment. Fig. 13-7 shows the converter with the moisture-resistant cover removed.

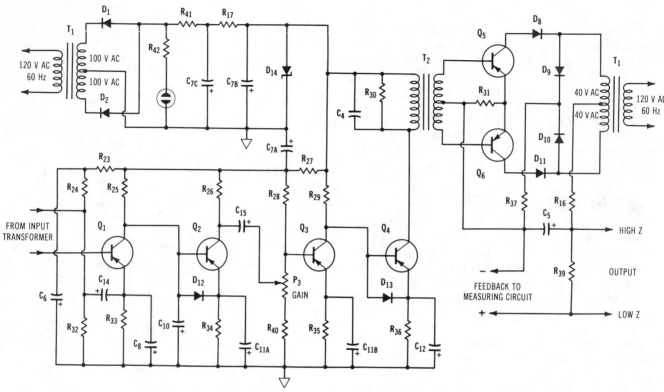

Fig. 13-5. Amplifier and demodulator circuits.

Courtesy Leeds & Northrup Co.

Fig. 13-6. An electric-to-pneumatic converter.

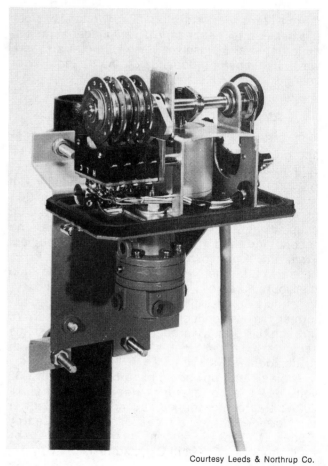

Courtesy Leeds & Northrup Co.

Fig. 13-7. Converter of Fig. 13-6 with cover removed.

SUMMARY

Converters are devices that form a link between a process transducer input and the controlled output. Typical converters have low-voltage inputs (millivolts) and outputs that are a value of current or pressure.

A millivolt-to-current (mV/I) converter employs a zener-diode-regulated reference voltage, measuring circuit, filter, dc-to-ac converter, amplifier, demodulator, and output circuitry.

The measuring circuit adds two voltages algebraically to the input millivolt signal. The first one is added to the input by the bridge circuit. Feedback voltage is then added to the positive input terminal. The resulting output is filtered and applied to the converter where it is changed from dc into ac.

The ac signal is then amplified by several common-emitter transistor stages. The output is a push-pull stage composed of two transistors. These transistors amplify the signal on alternate half-cycles which would normally develop a sine-wave output. By employing diodes in the output, the ac is rectified. As a result, when the base and collector voltages are in phase, the collector will conduct current. The amount of current depends on the amplitude of the base voltage. The pulsating dc is then filtered with the output being an average value of the combined collector current.

Millivolt-to-pressure (mV/P) converters serve as a link between electrically energized controller action and a pressure-actuated output. Typical converters of this type change varying dc signals into 3–15 lb/in² (20.7–103.4 kPa) air-pressure signals.

CHAPTER 14

Magnetic Amplifiers

GENERAL INFORMATION

In this and following chapters we will study Foxboro equipment that performs the same functions as equipment described in preceding chapters. Although a Foxboro amplifier may use a magnetic amplifier instead of the transistorized device found in a Honeywell system, we will find that most of the component amplifiers used are completely transistorized. Even in a magnetic amplifier, transistors, and diodes are important components. We will need to keep in mind the discussions of these components in previous chapters.

This chapter describes the operation of a magnetic amplifier. In order to do this, some preliminary knowledge of the saturable-core reactor is necessary. After a general discussion of these subjects, we will move to their operation in the actual equipment. Five topics will be covered: transducers, recorders and indicators, controllers, converters, and accessories.

SATURABLE-CORE REACTORS

Before we can readily understand the saturable-core reactor, we need to have firmly in mind some notions concerning the magnetic circuit. Let us consider a toroidal ring of ferromagnetic material with a coil of wire wound tightly and distributed evenly about it (Fig. 14-1). When there is a current through the coil, the resulting magnetic flux is confined almost entirely in the ring. Hence, this is the only magnetic path that needs to be considered. A magnetic-circuit problem then presents itself. To solve the problem two things must be considered: the flux-producing ability of the coil and the susceptibility of the ring material to magnetism. This problem is more exact for the ring than for any other magnetic circuit geometry. This is because of the almost complete uniformity of all significant variables throughout the ring. For example, the flux lines form concentric circles within the ring and the area of this field will be the same for

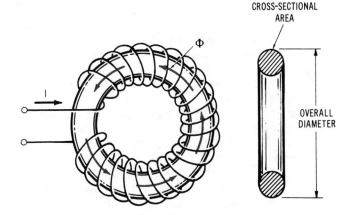

Fig. 14-1. Toroidal magnet.

any cross section. If the cross-sectional area of the ring is small compared to its overall diameter, then the flux path is approximately the same length anywhere. However, concepts developed from the ring are applicable to other magnetic-circuit geometries.

The flux-producing ability of the coil in Fig. 14-1 is proportional to the current through the coil and the number of turns. This is also true of any other magnetic circuit. The flux-producing ability of the coil is expressed as magnetomotive force ($\mathcal{F}$), and is given by the formula $\mathcal{F} = NI$ ampere-turns.

Besides the resulting flux being proportional to $\mathcal{F}$, it is also a function of the opposition of the core to carrying flux. This opposition is called the reluctance ($\mathcal{R}$) of the magnetic circuit. The flux Φ is then equal to $\Phi = \mathcal{F}/\mathcal{R}$ webers with $\mathcal{R}$ measured in ampere-turns per weber. As is true with resistance in the electric circuit, reluctance is proportional to the length of the path, inversely proportional to the cross-sectional area, and dependent on the material. The magnetic susceptibility of the core material is called its permeability, expressed as μ. Expressed quantitatively, $\mathcal{R} = 1/\mu A$. The equation, $\Phi = \mathcal{F}/\mathcal{R}$, is sometimes called the Ohm's law of the magnetic circuit and illustrates the similarity between the magnetic and the dc electric circuits. Table 14-1 may help to

Table 14-1. Comparison of Magnetic and Electrical Units

Magnetic-Circuit Quantity	Electric-Circuit Quantity
Magnetomotive Force ($\mathcal{F}$)	Electromotive Force (V)
Flux (Φ)	Current (I)
Reluctance ($\mathcal{R}$)	Resistance (R)
Flux Density (B)	Current Density (J)
Permeability (μ)	Conductivity (γ)

apply previously learned facts about the dc circuit to the magnetic circuit.

Kirchhoff's laws may be written for the magnetic circuit. The first, analogous to the current law, is that the sum of the flux entering a junction in a magnetic circuit equals the sum of the flux leaving that junction. The second, analogous to the voltage law, states that around any closed path in the magnetic circuit the algebraic sum of the magnetic potentials ($\Phi \times \mathcal{R}$ drops, analogous to the I $\times$ R drops in electric circuits) must equal the net ampere-turns ($\mathcal{F}$) of excitation.

Magnetic circuits, however, differ from electric circuits in one very important way. The reluctance of a magnetic-circuit path containing iron, or other ferromagnetic material, is a function of the flux in that path. As the flux increases, it takes a larger change in $\mathcal{F}$ to produce the same change in flux. The circuit is said to saturate. The fact that $\mathcal{R}$ and μ are not constant makes it unusual to use the formula, $\Phi = \mathcal{F}/\mathcal{R}$, as we use it in electric circuits. That is, direct number substitution is not usually performed with the formula since $\mathcal{R}$ varies with Φ. The main advantage lies in the ability to use the electric-circuit analogy in the thought process of solving the magnetic-circuit problem. A graphical method of finding these values is usually used because of these nonlinearities.

The graphical approach to the magnetic circuit makes use of a B-H magnetization curve (Fig. 14-2), where B is the flux density of the toroid and H is the magnetizing force, or the intensity of the magnetic field. In the toroid of Fig. 14-1, the flux (Φ) is considered to be evenly distributed over the cross-sectional area (A); therefore, the flux density (B) everywhere in the toroid is $B = \Phi/A$ teslas (webers per square meter). Because of the uniformity of the magnetic path of the toroid, the magnetomotive force ($\mathcal{F}$) expended per unit length

(*l*) of the toroid is constant. This quantity is referred to as the magnetizing force (H) and is given by the formula $H = \mathcal{F}/l$ ampere-turns per meter. By substitution, we can find the relationship between B and H as follows:

$$B = \frac{\Phi}{A},$$

but since

$$\Phi = \frac{\mathcal{F}}{\mathcal{R}},$$

then

$$B = \frac{\mathcal{F}}{\mathcal{R}A}.$$

Also,

$$\mathcal{R} = \frac{l}{\mu A},$$

and

$$\mathcal{F} = Hl,$$

therefore,

$$B = \frac{Hl}{(l/\mu A)A}$$
$$= \mu H.$$

This equation, $B = \mu H$, is a basic relationship involving the property of the magnetic material. The B-H curve depends only on the material of the toroid and not on its dimensions. This makes it possible to develop a table of universal curves for different materials that can easily be used to find circuit parameters when a given core is involved.

In Fig. 14-3, when an alternating current is applied to the coil, a magnetizing force results which varies from $+H_{max}$ to $-H_{max}$. This causes the flux density of the core material to vary from $+B_{max}$ to $-B_{max}$. At the start, however, when the current is

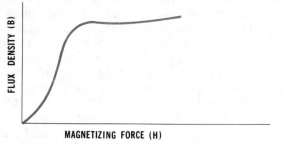

Fig. 14-2. A B-H magnetization curve.

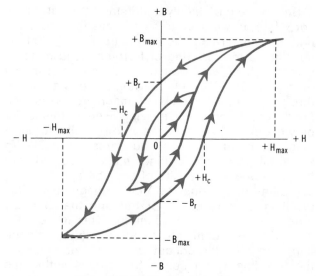

Fig. 14-3. B-H hysteresis loops.

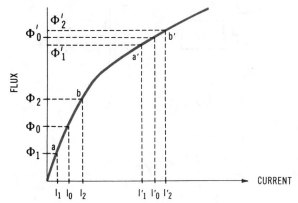

Fig. 14-4. Curve showing effect of changing the dc excitation point.

first applied and the core is unmagnetized, the curve starts at the zero point and increases to the saturation point at $+B_{max}$ as H values increase in a positive direction. H then decreases to zero, but B decreases only to the value of $+B_r$, not to zero, because of the circuit condition known as hysteresis. Hysteresis results from the fact that when the magnetizing force is removed, some residual magnetism remains in the core material. It takes a certain magnetizing force in the opposite direction to reduce the flux density to zero. This reverse magnetizing force ($-H_c$) is known as the coercive force of the core material. As H values then increase in a negative direction, B values also increase in a negative direction to the saturation point at $-B_{max}$. The $-H$ values then decrease to zero, B decreases to $-B_r$, and the loop is finally completed with positive H values causing saturation at $+B_{max}$ once again.

Thus, B varies with H about a loop known as a hysteresis loop (Fig. 14-3). For any given core material, the shape of this loop will always be the same. With a smaller ac current applied, however, a smaller loop results. Notice the similarity in shape of the two loops shown in Fig. 14-3.

Inductance (L) depends on the ratio of the change in flux to the change in current. Therefore,

the inductance of an iron-core coil can be varied by changing the portion of the magnetization curve or hysteresis loop being used. A variable reactor can be constructed using this principle. It employs a dc-excited coil to change the degree of saturation in the core on which the ac coils are wound.

In the magnetization curve shown in Fig. 14-4, if the dc excitation is constant at I_0, a superimposed ac excitation may cause variation from I_1 to I_2 and a flux change from Φ_1 to Φ_2. The apparent inductance to alternating current is then proportional to the slope of line ab. If, however, the dc excitation is changed to I'_0, the flux change for the same ac excitation is only from Φ'_1 to Φ'_2. In this region, the apparent inductance is proportional to the slope of line a'b', a much smaller value than before. A change in dc excitation has effectively changed the inductance of the ac system. Hysteresis effects are not included in this discussion.

The schematic of a common form of saturable-core reactor is shown in Fig. 14-5. The ac magnetomotive force acts on the magnetic path abdfeca. The dc magnetomotive force acts on paths cabdc and cefdc in parallel. The dc coil then controls the reactance of the ac circuit by adding flux to one of the outer legs of the core while subtracting flux from the other outer leg. The amounts being added and subtracted are unequal because of saturation. In other reactors, the ac and dc magnetomotive forces may be supplied by the same coil that carries an alternating current superimposed on a di-

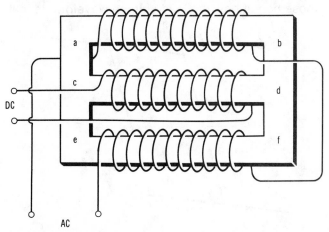

Fig. 14-5. Schematic of a saturable-core reactor.

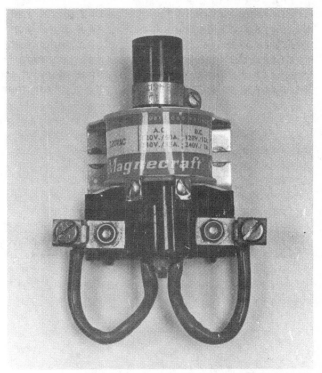

Fig. 14-6. A commercial saturable-core reactor.

rect current. Saturable-core reactors have many applications in control circuits. Special types will be used in magnetic amplifiers. Waveform distortion due to hysteresis and other core losses should be kept in mind as they are discussed. Fig. 14-6 shows a saturable-core reactor.

MAGNETIC AMPLIFIERS

A magnetic amplifier is a device that uses saturable-core reactors either alone or in combination with other circuit elements to secure amplification or control. The devices are amplifiers because the expenditure of relatively small amounts of power in the control winding permits control over relatively large amounts of power in the ouput windings. Magnetic amplifiers are also known by various trade names, such as Magamp and Amplistat.

A very common form of magnetic amplifier consists of saturable reactors in association with diode rectifiers. A specific example is given by the circuit diagram of Fig. 14-7. This is a single-phase, full-wave, center-tapped rectifier circuit. The magnitude of the direct current in the load is controlled by adjusting the relatively small current in the control windings. The principal source of power is the center-tapped transformer to the ac supply. A simplified explanation of the operation may be given by comparing the amplifier of Fig. 14-7 with the full-wave rectifier in Fig. 14-8A. Fig. 14-8A is simply a redrawing of the full-wave rectifier. The idealized load-voltage waveform is shown in Fig. 14-8B.

The magnetic amplifier of Fig. 14-7 differs from the simple rectifier circuit of Fig. 14-8 only by the addition of the saturable reactors in series with each rectifier element. The object of the reactors is to permit conduction by a rectifier element to be delayed after the positive half-cycle for that

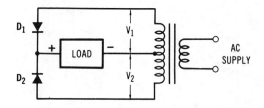

(A) Schematic diagram.

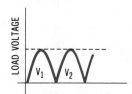

(B) Load-voltage waveform.

Fig. 14-8. Rectifier circuit.

element starts. The delay in the magnetic amplifier is determined by the properties of the magnetic core in conjunction with the value of the control magnetomotive force.

An idealized hysteresis loop for a typical core material is shown in Fig. 14-9. Until the core flux on the positive half-cycle of voltage reaches a value corresponding to the almost horizontal line def, the reactor presents such a high impedance that practically all the voltage is used up across it. Only a small amount of current is then present in the rectifier element and, hence, in the load. When the core becomes saturated, however, the slope of line def is so small that the impedance of the reactor is practically negligible. Practically all of the voltage appears across the load, and the rectifier is conducting a fairly large amount of current. On the negative half-cycle, diode D_1 blocks current through it, and diode D_2 undergoes the preceding process.

The point in the positive half-cycle at which saturation is reached can be controlled by the dc magnetomotive force of the control windings. Thus, if the control magnetomotive force corresponds to the abscissa of point d (Fig. 14-9) or any point to the right of it, the core is already saturated. Under these conditions, the load-voltage waveform is that

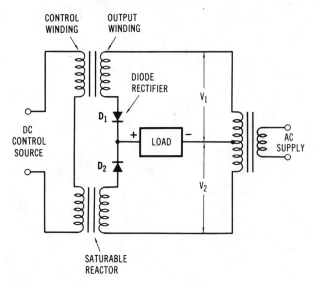

Fig. 14-7. Magnetic amplifier circuit.

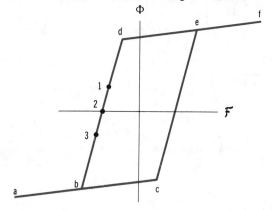

Fig. 14-9. Idealized hysteresis loop.

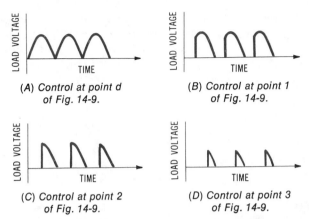

(A) Control at point d
of Fig. 14-9.

(B) Control at point 1
of Fig. 14-9.

(C) Control at point 2
of Fig. 14-9.

(D) Control at point 3
of Fig. 14-9.

Fig. 14-10. Load voltage of a magnetic amplifier.

of complete full-wave rectification as shown in Fig. 14-10A. If, however, the control magnetomotive force corresponds to the abscissa at point 1, time must elapse until the flux builds up to the saturation value before effective conduction can start. The load-voltage waveform becomes like that of Fig. 14-10B. For points farther to the left, such as 2 and 3, still further delay is occasioned. The resulting waveforms are approximately as in Figs. 14-10C and D. Finally, a value of control magnetomotive force will be reached when no effective conduction takes place and the load voltage becomes practically zero.

This magnetic amplifier, then, gives the same general type of control over the output current as is obtained by a grid-controlled rectifier tube called a thyratron, or a solid-state device known as a silicon controlled rectifier, or SCR.

The saturable reactors and diode rectifiers are quite rugged and require less maintenance than many other types of amplifiers. Additional windings may be added to provide for feedback to give better amplification or better control. Magnetic amplifiers are available in power sizes from a few milliwatts up to hundreds of watts. The development of magnetic amplifiers has to a considerable extent been dependent upon the development of suitable diode rectifiers and core materials with properties approaching those in Fig. 14-9.

Magnetic amplifiers are used in industry today to achieve dc power control. These applications range from simple self-contained units to large three-phase input devices that employ feedback control circuits. Since control of this type employs no moving parts, it serves as a very efficient and reliable method of controlling large amounts of power. Fig. 14-11 shows a general-purpose magnetic amplifier circuit application.

SUMMARY

A magnetic amplifier is primarily a saturable-core reactor employing diodes that permits control of large amounts of electrical power.

A saturable-core reactor has two or more ac windings connected in a series arrangement. Applying dc to an independent core winding increases or decreases the magnetomotive force of the core to a level such that large amounts of ac can be controlled by small values of dc. When saturation of the core occurs, coil inductance is changed or controlled as a result of this action.

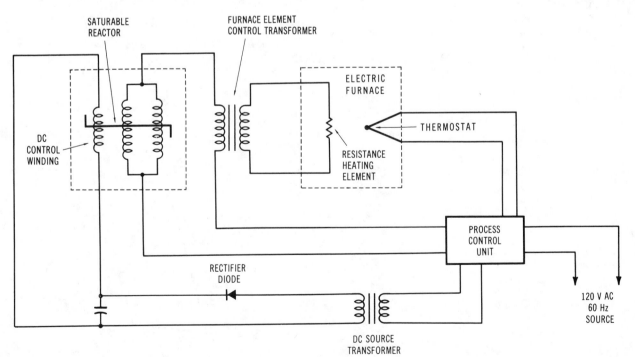

Fig. 14-11. General-purpose magnetic amplifier circuit application.

B-H curves are used to show the relationship between flux density and the magnetizing force of a coil. Inductance depends on the ratio of the change in flux to the change in current. The inductance of an iron-core coil can be altered by changing the portion of the magnetization curve or hysteresis loop being used. Through this action, coil reactance can be altered, which permits control of large amounts of ac power without moving parts.

CHAPTER 15

Magnetic Transmitters and Converters

GENERAL INFORMATION

Foxboro transmitters and converters are covered together in this chapter because a combination of all the components discussed here might be used to provide a single recorder or controller input. Input transmitters to be discussed involve a differential-pressure transmitter, an R/I converter, and a V/I converter. These components may individually provide a signal input to a recorder or controller. Combinations of these components, however, may feed into the following converters: a summing amplifier, a multiplier/divider amplifier, or a square-root converter. All of the foregoing components provide a 10–50-milliampere output. An additional converter will provide an output current range of 0–40 milliamperes for receiving components that require such an input.

DIFFERENTIAL-PRESSURE TRANSMITTER

The differential-pressure transmitter can be adjusted to measure differential pressures of 0–850 inches of water (0–210 kilopascals). Modifications can be made so that it will measure flow rate up to 30 cubic feet per minute (14 liters per second). It can also measure liquid level, static pressure, and absolute pressure. The transmitter is composed of three basic sections: the body, the force-balance system, and the oscillator/amplifier. The body consists of the component parts of the transmitter that come in contact with the fluids to be measured. The sensing element housed in this area is shown in Fig. 15-1. This element is composed of two metal diaphragms separated by a silicone fluid. This type of construction prevents overrange damage and provides for noise filtering at the point of measurement.

The force-balance system is composed of a detector and a feedback motor. Detection is provided by a movable laminated core (armature) in an inductor or coil. The armature position is varied by a lever system connected to the sensing element. The feedback motor provides a mechanical null on the lever connected to the armature, as shown in Fig. 15-1. The detector is excited by the amplifier and it also amplifies its output. The oscillator/amplifier is fully transistorized and is much like those used in Honeywell equipment discussed previously. The 10–50-milliampere output signal passes through the feedback motor as well as the load. From the diagram, it can be seen that a two-wire transmitting system is used. The oscillator/amplifier can be obtained both as a remote unit and as the integrally mounted unit shown here.

High- and low-pressure inputs are applied to the diaphragm capsule, as shown in Fig. 15-1. A change in the differential pressure across the diaphragm will cause a slight movement of the armature. As shown in the diagram, this movement is accomplished through a system of levers. The

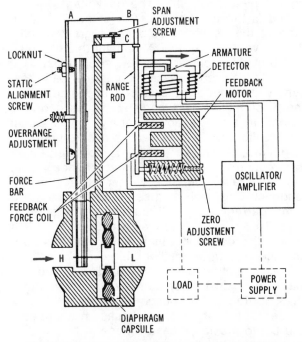

Fig. 15-1. Foxboro differential-pressure transmitter.

amount of movement of the armature is set by three adjustment screws: the zero adjustment, the range adjustment, and the overrange adjustment. An increase in high pressure decreases the air gap between the primary and secondary cores in the detector transformer. This movement increases the coupling between the two windings, and as a result the amplifier input increases.

The oscillator/amplifier used with the differential pressure transmitter is shown in Fig. 15-2. Transistor Q_1 acts as the oscillator; Q_2 and Q_3 form a two-stage dc-coupled Darlington amplifier. The detector signal is applied to the base of Q_1. The amplified output is taken from the collector. As described in the section on transistors, the kind of amplifier hookup provides an output that is in phase with the input as long as the load is resistive. The load of Q_1 is composed of the primary coil of the output transformer (T_1) for this stage, with a parallel capacitor (C_3). These components form a resonant circuit at about 1 kHz. It is, therefore, a resistive load at that frequency. Feedback coupling is provided by a capacitor (C_1) in series with the primary detector coil. These two components form a series-resonant circuit at 1 kHz. Feedback is in phase at a frequency of 1 kHz, so the circuit of Q_1 will oscillate at this frequency. The amplitude of this oscillation depends on the amount of coupling in the input detector transformer. This, of course, is determined by the position of the armature.

The oscillator output, the output of Q_1, is applied to the base of Q_2 by transformer coupling. The output of the coupling transformer is rectified by a full-wave, solid-state rectifier and filtered by an RC pi-section filter. The dc voltage thus obtained is the input to Q_2. The greater the amplitude of the oscillator output, the greater will be this dc voltage. Q_2 and Q_3 amplify this dc voltage and provide a current output proportional to the input. The current passes through the load, power supply, and feedback force coil, and is adjusted to the 10–50-milliampere range by a series load-adjust potentiometer. The power supply contains a full-wave solid-state rectifier and a choke input filter. As shown, the power supply is normally located in the control room with the load or receiver, and the two-wire transmission system is used. Since the feedback force coil is in series with the load, the same current passes through it as through the load. This current produces a magnetic field that moves the detector armature to keep the force-balance system in equilibrium.

You will notice that a zener diode (D_5) is used to maintain a constant emitter-to-collector voltage in Q_1. A portion of this voltage is taken from a voltage divider formed of a resistor (R_8) and two diodes (D_3 and D_4) to provide a base voltage to bias Q_1. This bias voltage is applied through the secondary of the detector coil. Since the bias voltage is applied through this coil, it will also bias the transformer. This bias will not change since it is zener-controlled and filtered by a parallel capacitor (C_6).

RESISTANCE-TO-CURRENT CONVERTER

Fig. 15-3 is a schematic diagram of the R/I converter used in the Foxboro equipment. It is composed of a four-stage magnetic amplifier using a resistance-bridge input. The same type of magnetic amplifier is used in the V/I converter, summing amplifier, and square-root converter. In succeeding discussions, it will be shown in block diagram form only. References will also be made to this diagram when discussing some of the controller circuits.

The four magnetic amplifiers used in the R/I converter are also identical to those outlined in the preceding material. The main difference between Figs. 15-3 and 14-7 is the addition of windings for feedback purposes. The addition of one or more dc bias windings does not vary the general operation of a magnetic amplifier. It merely means that the dc bias excitation will be an average of the dc bias of each of these windings. In the case of the R/I converter, the additional windings provide for an ac negative feedback. These windings are wound in such a direction as to be out of phase with the output. The negative feedback will then oppose the output signal, thereby decreasing it. Negative feedback provides for circuit stabilization in exactly the same way that it does in transistor circuits.

The variable resistance is included in the input bridge circuit of Fig. 15-3. The resistance will normally be a temperature-sensitive resistance bulb. Dc excitation for this bridge is provided by a zener-diode-regulated power supply. The variable resistance controls the amount of voltage applied to winding C of the first stage of the magnetic amplifier. Any change in the value of the resistance bulb changes the balance point of the bridge. A millivolt output results in a range of 2–7 millivolts. The dc excitation of winding C determines the operating point of the magnetic amplifier on the B-H curve, thereby varying its output. This output is developed across windings G_1 and G_2. Excitation for these windings comes from the power transformer. This output is rectified by D_3 and D_4 and filtered by C_3 and C_4. The dc output of the first stage is then fed to the second stage by winding C. Winding B of the second stage feeds some of the input signal of the second stage through B_1 and B_2 of the first stage. These windings provide the negative feedback mentioned previously. F_1 and F_2 of the second stage provide for some compensation for phase distortion. The third and fourth stages of amplification operate in the same man-

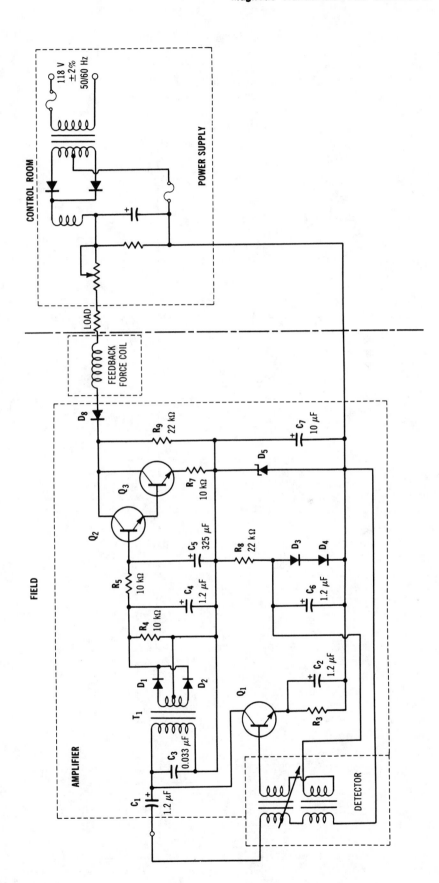

Fig. 15-2 Oscillator/amplifier.

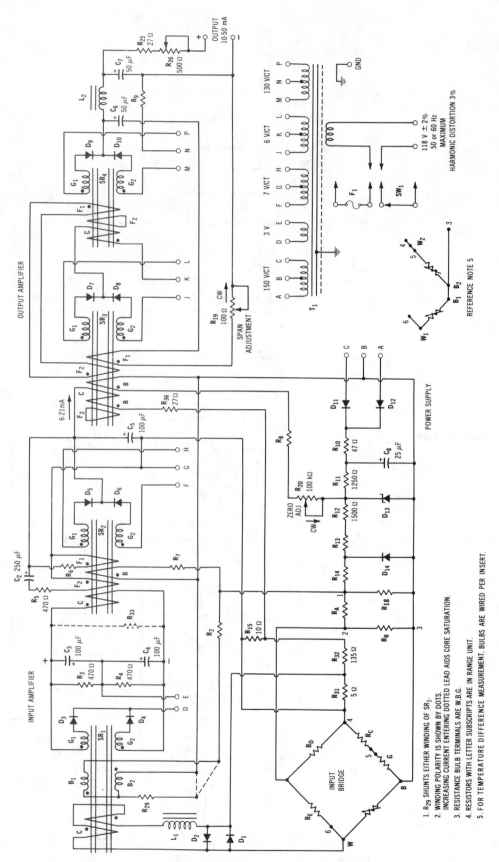

Fig. 15-3. Foxboro resistance-to-current converter.

ner as the other two. Winding F_1 of the fourth stage provides feedback to F_2 of the third stage. Feedback also occurs from the output of F_1 of the third stage. This feedback circuit involves the use of dc excitation. A dc currrent also passes through winding B of the third stage. This maintains the average bias point of that amplifier. The output of the fourth stage is filtered by L_2, C_6, and C_7. The dc output, adjusted by R_{26}, will be a 10–50-milliampere dc current that is proportional to the variable resistance in the input bridge.

VOLTAGE-TO-CURRENT CONVERTER

Fig. 15-4 is the block diagram of the V/I converter. The only circuit difference between this amplifier and the R/I converter is the method of input. A millivolt or thermocouple input is fed to the first stage of amplification in parallel with the input bridge. In our discussion, we will consider the bridge voltage to be stable, so that the only input will be from the millivolt source. This converter is normally used for thermocouple compensation or to establish the zero reference point for the input of a controller.

The span resistor circuit shown here involves R_2, R_7, and R_{29} (see Fig. 15-3). R_{29} can be connected in parallel with either winding B_1 or B_2 to obtain the amount of feedback desired. This feedback affects the gain and, therefore, the span of operation of the amplifier. Further span adjustment can be obtained by R_{19} in the feedback circuit between the third and fourth amplifiers. By adjusting the bias in the third stage of amplification by R_{20}, the zero reference point is set. Having a separate feedback system for the third and fourth amplifiers makes it possible to ground either thermocouple lead as well as to stabilize the circuit.

SUMMING AMPLIFIER

Fig. 15-5 is a schematic diagram of the summing amplifier. It is a one-stage magnetic amplifier much like the four used in the preceding converters. The primary difference between this and preceding magnetic amplifiers is the presence of eight input windings. All of these windings are wound in the same direction, and their dc voltage inputs should be connected in the same direction for summing to occur. The sum of the dc inputs of these eight windings determines the saturation of the core. The output, which is obtained in the manner previously described, will be proportional to the sum of these inputs. In order to get this output in the current range of 10–50 milliamperes, zero and span adjustments are used. The zero adjustment sets the current in the bias winding so an output of 10 milliamperes occurs with minimum inputs. The span adjustment determines the amount of negative feedback and, therefore, sets the gain of the amplifier. This adjustment will also affect the zero setting. The two adjustments will, therefore, have to be used together until the proper settings have been determined.

SQUARE-ROOT CONVERTER

Fig. 15-6 shows the block diagram of the square-root converter. The amplifier section of the converter is identical to the two-stage magnetic amplifiers used in preceding converters; the only difference is found in the feedback loop. Negative feedback is accomplished through a variable-resistance diode function generator. As the input to the amplifier increases, the gain of the amplifier is varied by automatically changing the resistance of this feedback loop. The resulting variation in

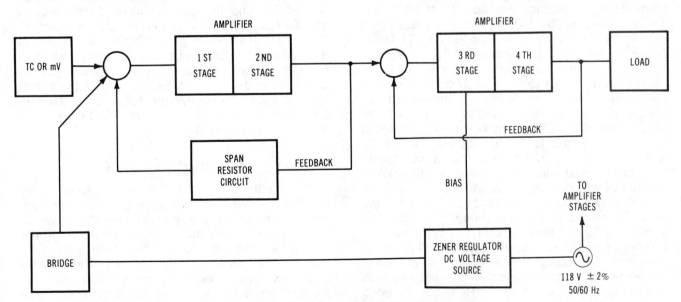

Fig. 15-4. Block diagram of voltage-to-current converter.

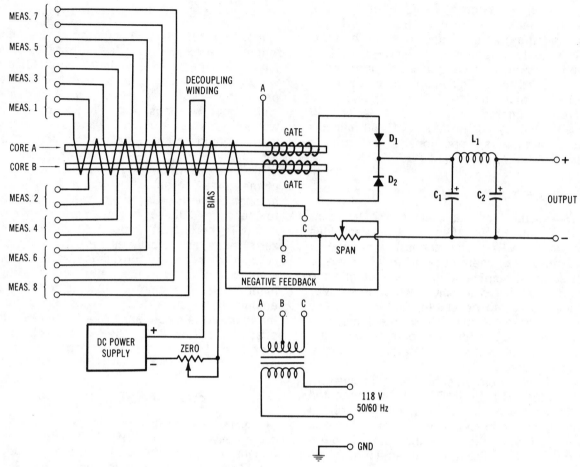

Fig. 15-5. Summing amplifier.

gain is such that the output is a square-root function of the input.

Fig. 15-6B schematically represents the feedback loop of the square-root converter. A portion of the amplifier output signal is fed back to the amplifier input through resistor R_1 at all times. This negative feedback decreases the overall gain of the amplifier. If any of the diodes (D_1, D_2, D_3, or D_4) are conducting, then resistors R_2, R_3, R_4, and R_5 are connected in parallel with R_1. This decreases the resistance in the feedback circuit, thereby increasing the feedback current. An increase in feedback current further decreases amplifier gain. When the input signal is very small, none of the diodes is conducting. As a result, R_1 is the only resistor in the feedback circuit. Amplifier gain is maximum under these conditions. As the input increases, diode D_1 will start to conduct. The point at which it conducts depends on the potential across it; i.e., the voltage from point X to point Y. The voltage at point Y is supplied by a zener-diode-regulated power supply and is developed across a voltage divider composed of R_6 and R_2. When the potential at point X is positive with respect to point Y, D_1 conducts. R_2 is then included in the feedback path in parallel with R_1. Feedback

current increases; the gain decreases. The potential at point Z, which is higher than that at point Y, is determined by voltage divider R_7 and R_3. When amplifier output at point X exceeds the value of point Z, D_2 conducts. Now R_3 is in parallel with R_1 and R_2. Amplifier gain is further decreased. D_3 and D_4 function in the same stepwise manner as D_1 and D_2. If we were to graph the output versus the input, we would find a close approximation to the square-root function. The fact that the resistance of the diode varies somewhat with the signal applied helps the converter accuracy. Maximum deviation from the square-root function occurs with small signal inputs.

MULTIPLIER/DIVIDER

The multiplier/divider will be discussed only functionally here. This is done because the converter is made up entirely of solid-state components and involves no magnetic-amplifier principles. It is described functionally here because it may be used as a receiver for other transmitters and converters covered in this section and may provide inputs for recorders and controllers to be discussed later. The block diagram in Fig. 15-7

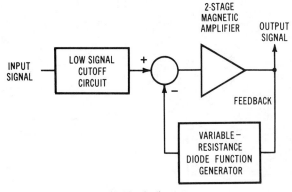

(A) Block diagram.

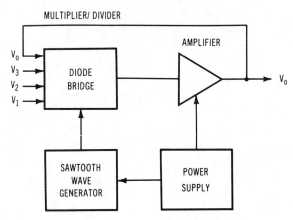

Fig. 15-7. Block diagram of multiplier/divider.

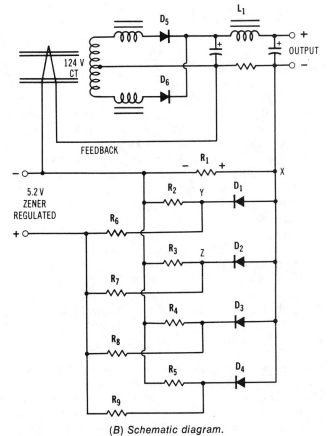

(B) Schematic diagram.

Fig. 15-6. Square-root converter.

indicates that the multiplier/divider converter can have three 10–50-milliampere inputs. These inputs (v_1, v_2, and v_3) will have certain computations performed depending on the equation that is to be solved. The circuitry needed to solve a particular equation will be installed by Foxboro. The type of diode-matrix bridge installed determines the computations involved. The bridge is triggered by a sawtooth generator, which delivers a signal such as that shown in Fig. 15-8.

The frequency of this sawtooth waveform is 10 kHz. The transient waveform is produced by an oscillator triggering a series RC circuit. It allows the capacitor to charge to only about 10% of its

maximum applied voltage, producing an almost linear charge rate. The amplitude of this waveform is not critical in this application. The output of the diode-matrix bridge is fed to an amplifier with a very high gain, producing an output in the range of 10–50 milliamperes proportional to the equations being solved. The multiplier converter can perform the following computations where v_1, v_2, and v_3 are the inputs, v_0 is the output, and k is a constant:

$$v_0 = \frac{v_1 v_2}{k}$$

$$v_0 = \frac{k v_1}{v_2}$$

$$v_0 = \frac{v_1 v_2}{v_3}$$

$$v_0 = v_1 v_2$$

$$v_0 = \frac{v_1^2}{k}$$

$$v_0 = k v_1$$

$$v_0 = \frac{v_1^2}{v_2}$$

CURRENT SOURCE

This unit is designed to take a 10–50-milliampere input and produce a 0–40-milliampere proportional current output. This is done for those receivers that need a zero reference voltage or current. Fig. 15-9 is a schematic diagram of the current source unit. The circuitry used in this converter is designed to produce a constant 10 milliamperes of

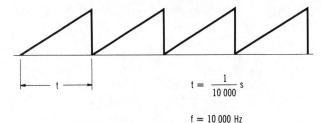

$$t = \frac{1}{10\,000}\,s$$

$$f = 10\,000\,\text{Hz}$$

Fig. 15-8. Sawtooth waveform.

171

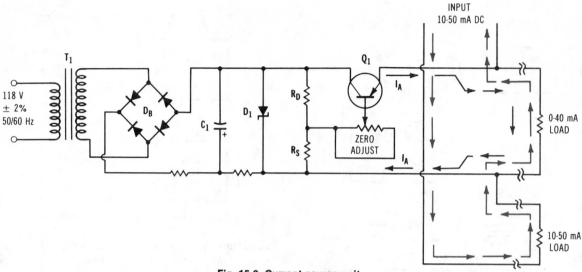

Fig. 15-9. Current source unit.

current over a wide range of temperature and loading conditions. The supply voltage is developed by bridge rectifier D_B and filter capacitor C_1. This power source is regulated by zener diode D_1. R_D and R_S act as a bleeder resistor and a voltage divider. A bleeder resistor helps to provide load regulation. The voltage divider also provides a base reference voltage or bias for Q_1. This reference voltage is adjustable. R_S is a temperature compensating resistor. Q_1 is an industrial grade silicon transistor, and, with the proper bias setting, will give a closely regulated 10 milliamperes. The current so produced (I_A) passes through the 0–40-milliampere load in the direction indicated. The input current of 10–50 milliamperes also passes through the 0–40-milliampere load and in the opposite direction of I_A. The two currents add algebraically in this load to produce the 0–40-milliampere output. I_A will not pass through the input transmitter due to its relatively high impedance.

SUMMARY

The magnetic transmitters and converters discussed in this chapter are designed to provide input for a recorder or controller. Inputs involve a differential-pressure transmitter, a resistance-to-current converter, and a voltage-to-current converter. The input signal is followed by summing amplifiers, multiplier/divider amplifiers, or square-root converter circuits. The resulting output is generally 10–50 milliamperes.

Differential-pressure transmitters are designed to measure pressures of 0–850 inches of water (0–210 kilopascals). They contain the body, force-balance system, and the oscillator/amplifier. The body deals with those things that come in contact with the fluids being measured. The force-balance unit employs a detector and a feedback motor. The oscillator/amplifier is transistorized.

An R/I converter contains a four-stage magnetic amplifier using a resistance bridge. Typically this amplifier is the same as the one used in the V/I converter, summing amplifier, and square-root converter. The major difference in the magnetic amplifiers of this unit and those of Chapter 14 is the feedback winding. Negative feedback provides for circuit stabilization.

V/I converters utilize a millivolt or thermocouple input to feed the first stage of amplification in parallel with the input bridge. Thermocouples are also used to compensate for temperature or to etsablish a zero reference for the input of the controller.

A magnetic summing amplifier has the possibility of eight input windings. All of these windings are wound in the same direction for summing to occur. The resulting sum of the winding inputs determines the saturation of the core. The output is proportional to the sum of the inputs with a current range of 10–50 milliamperes.

Square-root converters employ magnetic amplifiers that have negative feedback through a variable-resistance diode function generator. An increase in input automatically changes the resistance of the feedback loop. The gain is such that the output is a square-root function of the input.

A multiplier/divider is a solid-state circuit that can have three 10–50-milliampere inputs. It is designed to solve a number of particular computations. The diode matrix bridge installed by the manufacturer determines the computations the unit will perform.

CHAPTER 16

Magnetic Recorders and Indicators

GENERAL INFORMATION

The recorders and indicators used in Foxboro equipment are somewhat different from those discussed previously. In operation, they are much simpler than the others. In theory, they operate much the same as the D'Arsonval meter movement. This meter movement is used in almost all sensitive test instruments. It will be discussed as an introduction to the recorder/indicator mechanism. The Foxboro integrator unit is also discussed in this chapter. The reason for including the integrator is because it is mechanically linked to the recorder motor.

MODEL 64 RECORDER

Fig. 16-1A is a pictorial representation of the D'Arsonval meter movement. The similarity between this and the recorder motor can be seen by comparing Fig. 16-1A and Fig. 16-1B. You will notice in Fig. 16-1A that the primary operating components are a moving coil, or armature, and a permanent magnet field. When a current is applied to the armature in the direction shown, the left side of the armature becomes a north magnetic pole. This induced north pole is opposed by the north magnetic pole of the field. There is also some attraction between the north pole of the armature and the south pole of the permanent magnet. The attractive force adds very little force to the armature compared to the opposing force because of the difference in distance. The right end of the armature is affected by forces similar to those acting on the left end. These forces result in the armature rotating in a clockwise direction. The direction of rotation depends on the rest or zero position of the armature. As a result, the armature is placed at a slight angle with the magnetic lines of force of the field. The rotating forces acting on the armature are opposed by a spring. Since the amount of tension on an ideal spring is proportional to the force applied, the spring pro-

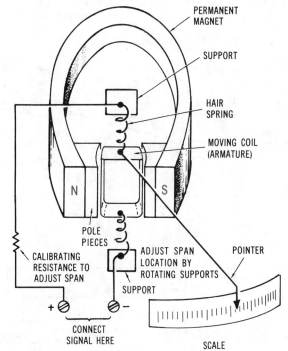

(A) D'Arsonval movement.

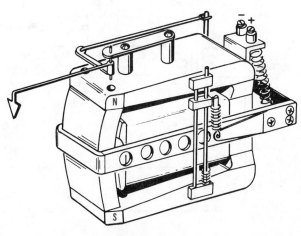

(B) Pen motor.

Fig. 16-1. Comparison of D'Arsonval meter movement and pen motor.

duces no nonlinearity in armature deflection. An increase in current increases the pole strength on the armature, and a greater deflection of the pointer results. The mechanical movement of the pointer is proportional to the current through the meter.

Fig. 16-1B is a pictorial representation of the recorder pen motor. A strong alnico magnet is used as the permanent field in this motor. A small moving coil (armature) is placed near one pole of this magnet. The motion of this coil is transferred by means of levers into recorder pen movement. A calibrating spring is used as the opposing force to armature movement. By adjusting the tension of this spring, the zero position of the pen can be set, and since the resistance of the armature is very small, several can be connected in series

without serious error. The recorder input is the standard 10–50 milliamperes delivered by Foxboro transmitters. Parallel capacitors act as noise filters.

MODEL 68 INTEGRATOR

The Model 68 integrator is a form of indicator used in conjunction with the Model 64 recorder and/or indicator. The integrator is composed of a cam switch that actuates a photodiode; a two-stage dc amplifier actuated by the photodiode; and a glass-enclosed reed switch, controlled by the amplifier, which starts or stops a counter motor. The counter can provide an indication that is proportional to the sum of the 10–50-milliampere input signal or a summation of the square root of

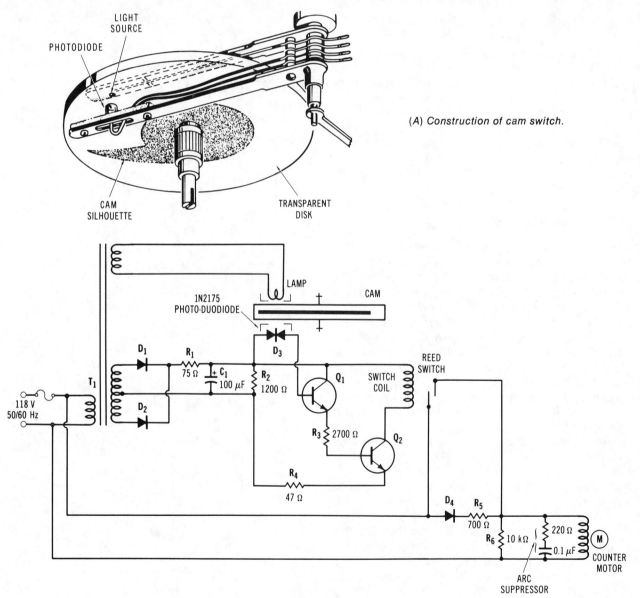

(A) Construction of cam switch.

(B) Schematic of integrator.

Fig. 16-2. Foxboro Model 68 integrator.

the signal. The type of indication provided is determined by the opaque portion of the cam switch and will not be discussed here. The square-root cam will total the flow rate from differential-pressure information.

Fig. 16-2A is a pictorial representation of the cam switch. A constantly rotating glass disk with an opaque cam shape near its center, which rotates between an actuating lamp and a photosensitive diode, makes up this switch. The actuating lamp above the glass disk is run below its rated value to assure it of long life. The lamp is positioned on a pivoted radial arm so that it is always directly above the photodiode, which is also positioned on this radial arm below the glass disk. The radial arm is positioned by a mechanical link with the recorder and/or indicator. As the indicator moves upscale, the radial arm will move away from the center of the rotating glass disk. With a cam such as the one shown in Fig. 16-2A, as the arm moves outward, the light from the lamp will shine on the photodiode for a longer period of time. When light strikes the photodiode, its resistance decreases.

Fig. 16-2B is a schematic diagram of the Model 68 integrator. Diodes D_1 and D_2 form a full-wave rectifier for the power supply of the two-stage amplifier. R_1 and C_1 are its filter, while R_2 acts as a bleeder resistor. D_3, the photosensitive diode, is in the base circuit of transistor Q_1, and controls its bias. Changes of resistance of D_3 change the current through Q_1, which in turn changes the current through Q_2. When light strikes D_3, enough current conducts through transistor Q_2 to actuate the relay in its collector circuit. This solenoid closes the glass-enclosed reed switch. The switch will remain closed until the opaque cam breaks the light beam between the actuating lamp and the photodiode.

When the reed switch is closed, 118 volts ac is applied to the counter motor. This ac motor will run under these conditions. The filter networks in parallel with the motor provide for arc suppression across the reed switch and for waveform improvement. When the reed switch is open, diode D_4 and resistor R_5 are placed in series with the motor. These components act as an electronic brake for the motor and provide for almost instant stopping of the motor. The rectifier applies a dc voltage to the ac motor to perform this function. So the actuating lamp shining on the photodiode through the transparent portion of the cam switch decreases the resistance of the photocell. The current through the two-stage dc amplifier increases. The switch relay is energized. The switch shorts out diode D_4 and provides ac voltage to run the counter motor. When the opaque portion of the cam switch breaks the light beam, the counter motor stops immediately.

A THERMAL-WRITING RECORDING INSTRUMENT

Thermal writing instruments use the magnetic movement principle to move the position of a writing stylus as a chart moves with respect to time.

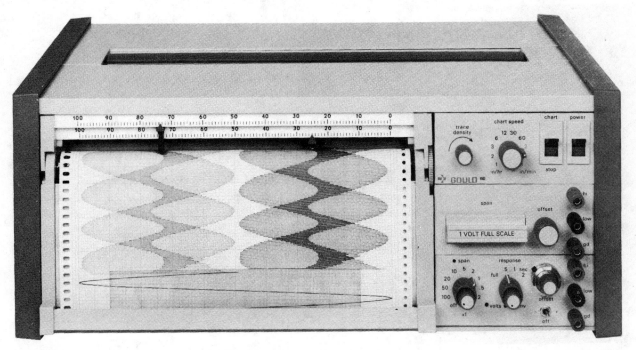

Courtesy Gould, Inc., Instrument Systems Division

Fig. 16-3. A thermal-writing strip-chart recorder.

The stylus, in this case, is heated instead of making an ink mark on the chart. Heat causes a permanent mark to be scribed into the temperature-sensitive chart paper. In practice, this recording principle is better than the ink principle because it eliminates problems of ink smudges, clogged ink lines, and fountain filling operations. Fig. 16-3 shows a representative thermal-writing strip-chart recorder. Note the fine detail of the reproduced chart pattern.

SUMMARY

Many recorders and indicators in use today operate much the same as the D'Arsonval meter movement. This type of instrument is much simpler than the servomotor units. The D'Arsonval movement is used to drive the pen mechanism.

A D'Arsonval movement employs a permanent magnet and a movable coil attached to an indicator hand or pen mechanism. When the polarity of the movable coil is the same as the permanent magnet poles, repulsion of the movable coil occurs.

An integrator is normally used in conjunction with a D'Arsonval pen unit. It is composed of a cam switch that actuates a photodiode, a two-stage dc amplifier actuated by the photodiode, and a reed switch controlled by an amplifier which starts or stops a counter motor. A rotating disk with an opaque cam near its center rotates between the lamp and a photosensitive diode. When the reed switch is closed, 118 volts ac is applied to the counter motor. When the reed switch is open, a diode and resistor are placed in series with the motor. This serves as an electronic brake for the motor and provides rapid stopping action.

CHAPTER 17

Magnetic Controllers and Transducers

GENERAL INFORMATION

Three controller functions and one transducer are discussed in this chapter. The flow controller contains magnetic amplifiers and pertains to the subject being discussed. The universal controller is transistorized and is much the same in operation as the controllers covered previously. The batch controller involves the insertion of a switching unit in either the flow controller or the universal controller. The transducer to be discussed is a current-to-air transducer.

FLOW CONTROLLER

Fig. 17-1 is a schematic diagram of the Foxboro Model 61 flow controller function generator. This is a two-stage magnetic amplifier constructed and operated in the same manner as those covered in preceding chapters. Negative feedback is fed through a one-stage magnetic amplifier. This amplifier provides for isolation between the input and output circuits. (The reliability of magnetic amplifiers makes them particularly useful in the controller circuit whenever long time constants are required.) An additional advantage of the magnetic amplifier in this situation is the fact that it is basically a current amplifier, since a current output is desired.

After recognizing the type of magnetic amplifiers being used, let us look at a simplified block diagram, Fig. 17-2. The difference between a measured voltage, and a set-point voltage is the amplifier input. The two-stage magnetic amplifier amplifies this error signal to a 10–50-milliampere output. The current develops a voltage across a resistor in the output circuit. This voltage is the input to the feedback amplifier. Also incorporated in the input circuit of the feedback amplifier is an adjustable RC time constant that determines the reset time of the amplifier. The output of the feedback amplifier is developed across a potentiometer that functions as the proportional-band control

that effectively sets the gain of the feedback amplifier. This, of course, determines the amount of negative feedback, hence, the gain of the entire circuit. Auxiliary power supplies provide a voltage for manual set-point adjustment as well as manual regulation.

UNIVERSAL CONTROLLER

Fig. 17-3 is a schematic diagram of the universal controller. It performs the same functions as the flow controller, with the addition of a derivative time adjustment. Because of the long time constant of the derivative time adjustment, a transistorized amplifier is used. A complicated, bulky magnetic amplifier would be necessary to handle the long time constants necessary for the ratio time set mode controller. This controller operates in essentially the same manner as the flow controller. Adjusting the derivative resistor varies the time it takes the derivative capacitor to charge. This varies the time it takes the controller to react to an output change.

The input of the universal controller is developed in the same way as in the flow controller. This error signal is, however, fed into a capacitance bridge. Two capacitors of this bridge (C_{29} and C_{30}) are called varicap diodes. They are reverse biased and are operated on the portion of their operating curve where they will not conduct current. Therefore, they behave as small variable capacitances of just a few picofarads.

The capacitance bridge is excited by a 100-kHz signal developed by oscillation of the amplifier. The reactance of the bridge is changed by the application of a dc error signal. This changes the amplitude of the output of the four-stage amplifier. The negative feedback of the amplifier is transformer coupled from the collector circuit of the output transistor to the capacitance bridge. This signal is rectified so that a dc feedback signal is utilized. Manual set-point adjustment and regulation are identical to the flow controller.

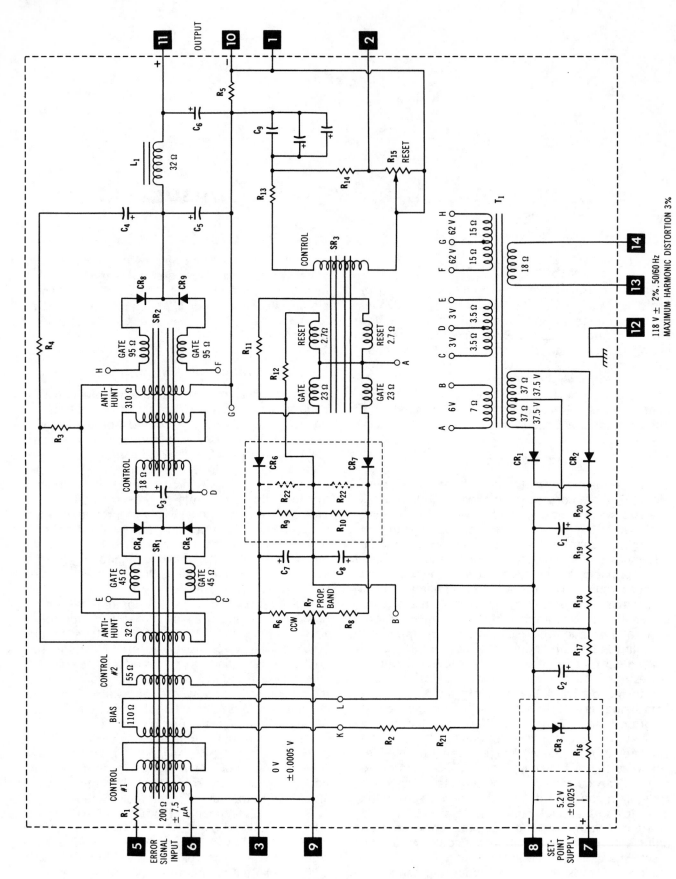

Fig. 17-1. Foxboro Model 61 flow controller function generator.

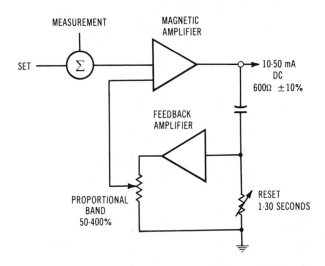

Fig. 17-2. Simplified block diagram of flow controller.

BATCH CONTROLLER

A block diagram of the batch controller is shown in Fig. 17-4. The batch controller is composed of either a Model 61 or Model 62 controller plus the batch control unit, which is simply a switching unit. With a normal controller, the error-signal input is the difference of the measured and set-point voltages, plus the feedback signal. If this signal exceeds ±10 millivolts the amplifier output is a full scale 10 or 50 milliamperes. If this condition exists for only a short period of time, the feedback capacitor becomes fully charged and the feedback signal will become zero. The output then will continue to be maximum.

The condition just described will be more likely to occur on batch or discontinuous processes. The batch control unit is inserted into the circuit to prevent overshoot when this condition exists. A diode limiting circuit in the batch control unit is triggered when amplifier output reaches a maximum. The batch unit then inserts a voltage into the feedback circuit which allows the capacitor to discharge. Discharge current is labeled "i" in the diagram. This current develops a feedback voltage across the proportional-band control. After capacitor discharge, the batch unit will switch out of the circuit. The measured value will then be able to approach the set point without overshooting.

CURRENT-TO-AIR TRANSDUCER

The final output of a controller is often converted to an air signal. The I/P transducer changes the standard 10–50-milliampere dc output of the controller to a standard 3–15-lb/in² (20–103 kPa) pressure output. Fig. 17-5 is a simplified diagram of the Foxboro I/P transducer. The dc controller output passes through the armature of the transducer torque motor. The magnetic force exerted

by the armature poles is proportional to this current. To explain, we will assume that there is a current increase. This increases the strength of both poles of the armature, as indicated in Fig. 17-5. The south magnetic pole of the armature is attracted to the north pole of the permanent magnet. The north pole of the armature is attracted to the south pole of the permanent magnet.

These two motions oppose each other. However, the south pole of the armature has a greater moment about the pivot (longer force arm) than the north pole, so that the resulting motion will be clockwise. This clockwise motion covers the nozzle, increasing the output pressure. This increase in pressure is fed back to the armature through the feedback bellows, thereby balancing the system. A 10–50-milliampere direct current produces a proportional 3–15-lb/in² (20–103 kPa) air signal.

SUMMARY

Magnetic amplifiers are commonly used in flow controller applications today. Typically this type of circuit employs two magnetic amplifiers. A summing amplifier receives the set-point and measured input signals which are combined and applied to one input of a magnetic amplifier. A second input to the same amplifier is from the negative feedback. The resulting output is 10 to 50 milliamperes dc. A sample of this is fed back to a second magnetic amplifier and used to develop the negative feedback signal. The output of the feedback amplifier is developed across a potentiometer that functions as a proportional-band control that sets the gain of the feedback signal.

Universal controllers perform the same function as the flow controller with the addition of a derivative time adjustment. Due to the long time constant of the derivative time adjustment, a transistorized amplifier is used. Adjustment of the derivative resistor varies the time it takes the derivative capacitor to charge. This determines the time that it takes the controller to react to an output change.

A batch controller is composed of either a flow or a universal controller plus a batch unit. The batch unit is basically a switching circuit. With a normal controller, the error-signal input is the difference between the measured and set-point voltages plus the feedback signal. If the signal exceeds ±10 millivolts, the output will go to its full-scale current indication. If this continues for a short period of time, the feedback capacitor becomes fully charged and the feedback signal goes to zero. The output will then continue at maximum. This would occur if the process is discontinued. The batch unit is, therefore, used to prevent overshoot when this condition exists.

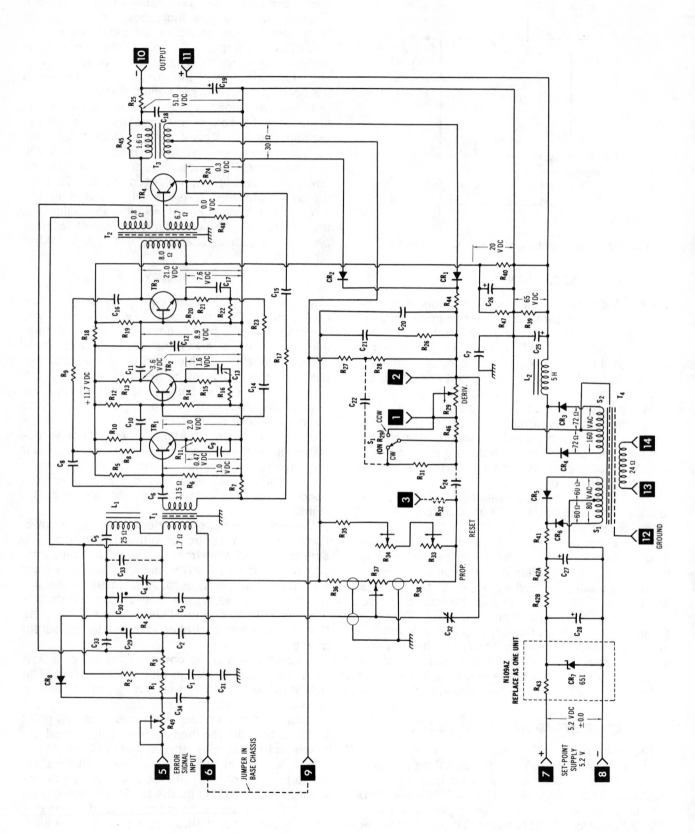

Fig. 17-3. Foxboro Model 62 universal controller function generator.

I/P transducers are used to change 10–50 milliamperes dc output to 3–15 lb/in² (20–103 kPa) pressure output. This is achieved by passing dc through the armature of a torque motor. Pivoting action of the armature controls nozzle pressure through a reducing tube and feedback bellows.

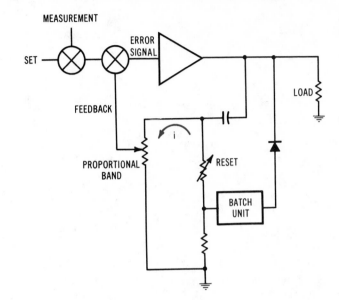

Fig. 17-4. Block diagram of batch control unit.

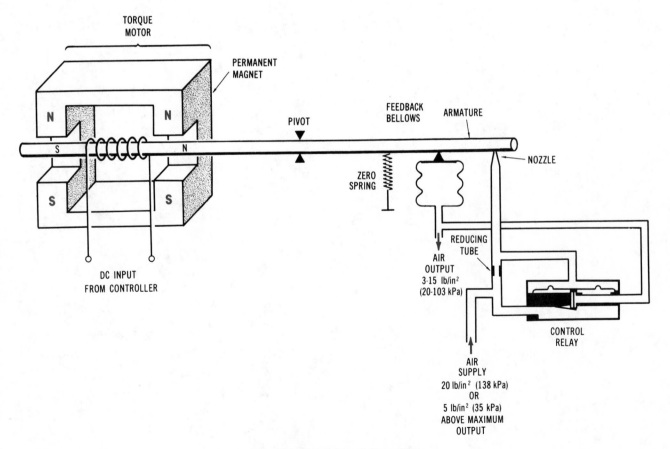

Fig. 17-5. Current-to-air transducer.

CHAPTER 18

Miscellaneous Accessories

GENERAL INFORMATION

Some of the accessories used in the Foxboro control system are discussed here. They are electronically controlled units that might need some explanation and that are pertinent to complete system operation. We will discuss the autoselector, ratio-set station, set-point station, and alarms.

AUTOSELECTOR CONTROL STATIONS

The autoselector permits the use of several controllers to operate a single output load such as a valve. Each of these controllers monitors related variables. In the situation chosen for demonstration of the autoselector, two controllers are feeding the autoselector. We want the one with the highest

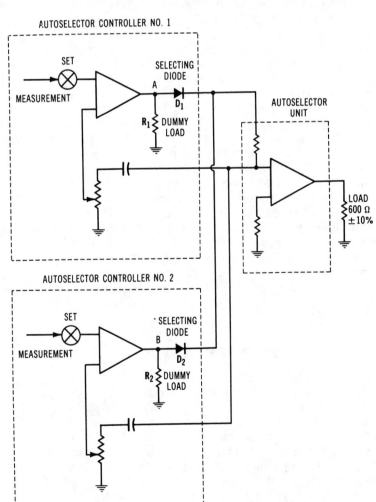

Fig. 18-1. Foxboro autoselector controller.

output to operate the valve, as shown in Fig. 18-1. Other modes of operation are available with the use of the autoselector; the situation that we illustrate here merely demonstrates its operation.

The number 1 controller in Fig. 18-1 is assumed to be delivering a 40-milliampere signal. The current develops a voltage drop across dummy load R_1. We will assume that the output of controller number 2 is 10 milliamperes. This current develops a voltage across R_2. R_1 and R_2 have equal resistances. The voltage at point A will then be higher than the voltage at point B. The voltages thus produced are such that diode D_1 will conduct and D_2 will not. The signal fed to the autoselector is then coming from controller number 1. If the output of controller number 2 increases, the voltage drop across R_2 increases. When the voltage drop across R_2 exceeds the voltage drop across R_1, D_2 will conduct and D_1 will not. The autoselector current then comes from the second controller.

RATIO-SET STATION

It is often desirable to use two controllers whose outputs maintain a constant ratio. For example, controller number 1 will have a variable output and the second controller output should be twice its value. In order to perform such a function a ratio-set station, represented schematically in Fig. 18-2, is used. The output of controller number 1 is represented by the letter A. This current develops a voltage drop across the 100-ohm resistor. At 0% signal (10 milliamperes dc) a 1-volt signal is developed. A zener-diode-regulated voltage source develops a 1-volt signal at point B. Since the polarities of these two voltages oppose each other, the potential between points X and Y is 0 volts. The magnetic amplifier is so biased as to produce a 10-milliampere output. As the signal at point A increases, a current passes through variable resistor R and the control winding of the magnetic amplifier. The setting of R determines the current through the control winding; R can be set so that the amplifier output is the desired ratio of the input at A.

SET-POINT STATION

The set-point station is a computer link with a process controller. The computer input to the set-point station is either an up-scale or a down-scale pulse. These pulses operate two relays. An up-scale pulse will close the relay, which provides a voltage to a reversible motor in such a direction as to drive the set-point control up-scale. When the desired set point is reached, the relay will be de-energized. The set-point control so operated is much the same as the manual set point previously mentioned. A down-scale pulse closes the second

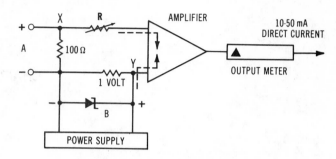

Fig. 18-2. Ratio-set station.

relay which provides a voltage to reverse the motor and drive the set point down-scale. The motor also drives a dial indicator and a feedback potentiometer. The voltage of the feedback potentiometer feeds a signal to the computer that corresponds to the input of the controller.

ALARMS

Foxboro alarm units are designed to open or close contacts when a measurement signal exceeds a preset limit. Although units are available for duplex alarms, only a single alarm will be discussed here. A simple wiring change will cause the alarm to function on a high signal rather than a low signal as demonstrated here.

Fig. 18-3 shows a simplified schematic diagram of a single alarm unit. Three voltage sources provide the input to a two-stage transistor amplifier. When a pulsating current passes through the primary of the coupling transformer T_X, it induces a voltage in the transformer secondary. The amplifier strengthens this ac signal to a level sufficient to trigger the alarm relay.

Two of the three inputs to the alarm amplifier are dc voltages and the other is a pulsating or ac voltage. The first of these voltages (V_i) is developed across resistor R_i by transmitter or controller output. This current is in the 10–50-milliampere range. The second dc voltage, V_8, is developed across resistor R_8 by a zener-regulated power sup-

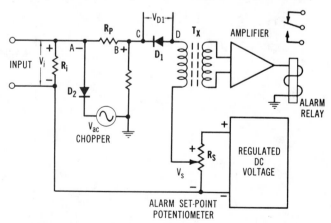

Fig. 18-3. Alarm unit.

ply. The amount of this voltage appearing in the alarm unit input is determined by the limit control, potentiometer R_s. The polarity of V_s opposes that of V_i. The voltage across diode D_1 is the difference between these two dc voltages. The diode will conduct only when the amplitude of V_s is greater than that of V_i.

The alternating voltage source, V_{ac}, is rectified by diode D_2. This half-wave rectified voltage is developed across resistor R_p. Since the ac voltage is rectified it will always have the same polarity. The polarity of this voltage is in such a direction as to add to V_i; therefore, it will further subtract from V_s. If diode D_1 is reverse biased (nonconducting) by V_i and V_s, V_{ac} will simply strengthen the reverse bias during each pulse. If, however, D_1 is forward biased (V_s greater than V_i), the V_{ac} will decrease the amount of forward bias during each half-cycle. This will produce a changing current through T_x, and an amplifier input will trigger the alarm.

The only wiring change necessary for a high-limit alarm is to change the polarity of diodes D_1 and D_2. Changing the polarity of D_1 means that it will conduct when V_i exceeds V_s. Changing D_2 will change the polarity of the half-wave pulses across R_p. Therefore, when D_1 is forward biased (V_i greater than V_s), V_{ac} will subtract from this forward bias. The resulting changes in current through T_x will again sound the alarm.

SUMMARY

Control systems have a number of accessories that are frequently used. This includes the auto-selector, ratio-set station, set-point station, and alarms.

An autoselector permits the use of several controllers to operate a single output load. Each controller deals with a specific variable. As a rule, the controller with the highest output is selected to perform the control function automatically.

The ratio-set station is designed for two controllers whose outputs are maintained at a constant ratio. A zener-diode-regulated voltage develops a signal across one output. A variable resistor is adjusted to produce a desired voltage ratio for the second output.

A set-point station serves as a computer link for a process controller. The computer input produces either an up-scale or down-scale pulse. These pulses will close a relay that provides voltage to a reversible motor which alters the output to coincide with the set-point position. When the set point is reached, the relay becomes de-energized.

An alarm is designed to open or close a set of contact points when a measurement signal exceeds a preset limit. Three voltage sources are supplied to this circuit. When a pulsating voltage is applied to the transformer primary, a voltage will be induced in the secondary, be amplified, and trigger the alarm. If a diode is reverse biased by the input voltage and set-point voltage no ac signal will pass to the transformer for alarm actuation. When the same diode becomes forward biased, a rectified ac will pass through the transformer and trigger the alarm. The set-point alarm voltage is adjustable to a desired value.

Electrical and Electronic Symbols

FIXED RESISTOR		COIL (AIR CORE)		COLD CATHODE	
TAPPED RESISTOR		COIL (MAGNETIC CORE)		PHOTOCATHODE	
VARIABLE RESISTOR		COIL (TAPPED)		POOL CATHODE	
RHEOSTAT		COIL (ADJUSTABLE)		IONICALLY HEATED CATHODE	
THERMISTOR		TRANSFORMER (AIR CORE)		GRID	
FIXED CAPACITOR				DEFLECTING ELECTRODE	
VARIABLE CAPACITOR		TRANSFORMER (MAGNETIC CORE)		ANODE OR PLATE	
POLARIZED CAPACITOR				TARGET OR X-RAY ANODE	
TURNSTILE ANTENNA		AUTOTRANSFORMER		DYNODE	
DIPOLE ANTENNA				IGNITOR OR STARTER	
LOOP ANTENNA		CURRENT TRANSFORMER		VACUUM-TYPE DIODE	
		RELAY, TRANSFER CONTACTS			
BATTERY				GAS-FILLED DIODE	
GENERAL ALTERNATING CURRENT SOURCE		**ELECTRON-TUBE SYMBOLS:**			
PERMANENT MAGNET		VACUUM-TYPE ENVELOPE		COLD-CATHODE, GAS-FILLED DIODE	
PIEZOELECTRIC CRYSTAL		GAS-FILLED ENVELOPE		VACUUM-TYPE PHOTOTUBE	
THERMOCOUPLE		FILAMENT AND DIRECTLY HEATED CATHODE		MULTIPLIER-TYPE PHOTOTUBE	
THERMAL CUTOUT		INDIRECTLY HEATED CATHODE			

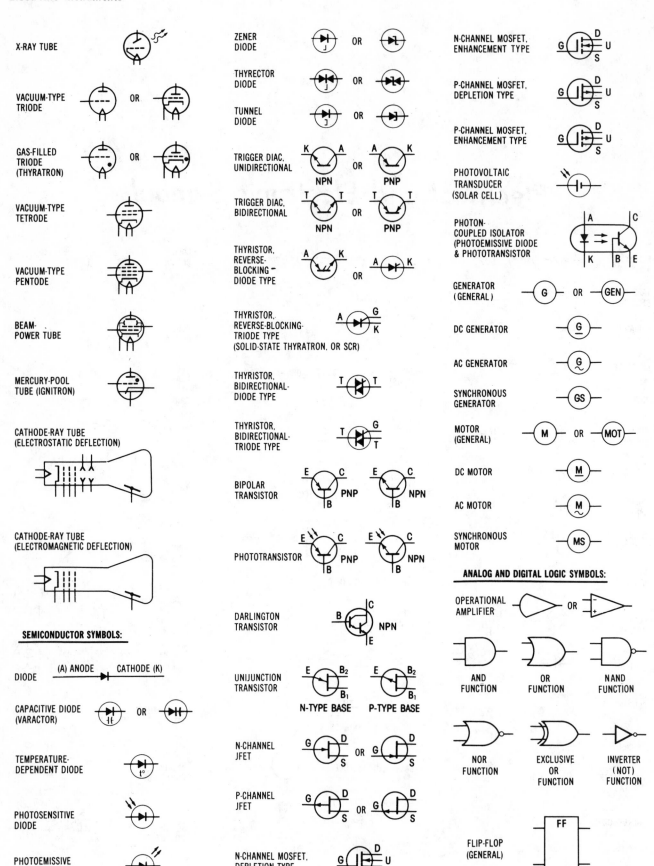

SEMICONDUCTOR SYMBOLS:

DIODE

CAPACITIVE DIODE (VARACTOR)

TEMPERATURE-DEPENDENT DIODE

PHOTOSENSITIVE DIODE

PHOTOEMISSIVE DIODE

ZENER DIODE

THYRECTOR DIODE

TUNNEL DIODE

TRIGGER DIAC, UNIDIRECTIONAL

TRIGGER DIAC, BIDIRECTIONAL

THYRISTOR, REVERSE-BLOCKING – DIODE TYPE

THYRISTOR, REVERSE-BLOCKING-TRIODE TYPE (SOLID-STATE THYRATRON, OR SCR)

THYRISTOR, BIDIRECTIONAL-DIODE TYPE

THYRISTOR, BIDIRECTIONAL-TRIODE TYPE

BIPOLAR TRANSISTOR

PHOTOTRANSISTOR

DARLINGTON TRANSISTOR

UNIJUNCTION TRANSISTOR

N-CHANNEL JFET

P-CHANNEL JFET

N-CHANNEL MOSFET, DEPLETION TYPE

N-CHANNEL MOSFET, ENHANCEMENT TYPE

P-CHANNEL MOSFET, DEPLETION TYPE

P-CHANNEL MOSFET, ENHANCEMENT TYPE

PHOTOVOLTAIC TRANSDUCER (SOLAR CELL)

PHOTON-COUPLED ISOLATOR (PHOTOEMISSIVE DIODE & PHOTOTRANSISTOR)

GENERATOR (GENERAL)

DC GENERATOR

AC GENERATOR

SYNCHRONOUS GENERATOR

MOTOR (GENERAL)

DC MOTOR

AC MOTOR

SYNCHRONOUS MOTOR

ANALOG AND DIGITAL LOGIC SYMBOLS:

OPERATIONAL AMPLIFIER

AND FUNCTION

OR FUNCTION

NAND FUNCTION

NOR FUNCTION

EXCLUSIVE OR FUNCTION

INVERTER (NOT) FUNCTION

FLIP-FLOP (GENERAL)

X-RAY TUBE

VACUUM-TYPE TRIODE

GAS-FILLED TRIODE (THYRATRON)

VACUUM-TYPE TETRODE

VACUUM-TYPE PENTODE

BEAM-POWER TUBE

MERCURY-POOL TUBE (IGNITRON)

CATHODE-RAY TUBE (ELECTROSTATIC DEFLECTION)

CATHODE-RAY TUBE (ELECTROMAGNETIC DEFLECTION)

Index